Tools of the Scribe

Brian Roark · Richard Sproat · Su-Youn Yoon

Tools of the Scribe

How Writing Systems, Technology,
and Human Factors Interact to Affect the Act
of Writing

 Springer

Brian Roark
Google
Portland, OR, USA

Richard Sproat ⓘ
Sakana.ai
Minato City, Tokyo, Japan

Su-Youn Yoon
Automated Scoring Team
EduLab Inc.
Minato City, Tokyo, Japan

ISBN 978-3-032-00830-5   ISBN 978-3-032-00831-2 (eBook)
https://doi.org/10.1007/978-3-032-00831-2

This Springer imprint is published by the registered company Springer Nature Switzerland AG
The registered company address is: Gewerbestrasse 11, 6330 Cham, Switzerland

If disposing of this product, please recycle the paper.

𒁾𒊬𒋗𒅗𒋫𒁲𒀀𒂊𒉈𒀫
𒁾𒊬𒊏𒀫

dub-sar šu ka-ta sa$_2$-a e-ne-am$_3$
dub-sar-ra-am$_3$

"A scribe whose hand can keep up with the mouth is indeed a scribe."

Sumerian Proverb Collection 2, quoted in Paulus (2023), p. 40.

Preface

The theme of this book is writing, and how the act of writing is influenced by the properties of a language's script and writing system, the writing devices one is using, one's own abilities in the language and writing system, and one's ability to engage in the gestures needed to write.

The writer has one main goal in mind: transfer his or her thoughts to a physical or virtual surface in a form that is visible (or in the case of Braille, tactile). How easy or hard this is depends on many factors. How well the writer knows the language and the writing system they are using. Whether the writer is physically disabled in any way. The complexity of the script or writing system that they must master in order to write. What particular writing technology is being used, since depending on what that technology is, and what script/writing system is assumed, that technological choice may make the task easier or harder. To give a simple example whose ilk we will expand upon at length in the course of this work, it is much easier to type on a traditional typewriter in English, than it would be to type in Kannada, or in Korean.

For many years the three of us have done research in areas that have circled around the topics we cover in this book. The first author has had a long-standing interest in input systems including smartphone keyboards for complex scripts such as the scripts of India, as well as input methods for people with motor and cognitive disabilities. The second author has a long-standing interest in scripts and writing systems, in particular how different scripts make use of the two dimensions afforded by a writing surface, what linguistic information writing systems represent, and how they represent it. The third author has many years' experience working on assessment methodologies for second-language learners and is actively involved in the application of Generative AI technology to the task of rating and giving feedback to students who are trying to master a second language.

Hence the current work, wherein we explore the various dimensions that affect the act of writing. While there have been many works on writing systems, second language writing, typewriters, modern computer-based input systems, and so forth, we are unaware of any other work that considers in one place all of the various dimensions that make writing easier or harder. We therefore believe, and hope, that this book will provide a unique perspective on a technology—writing—that has been with us in one way or another for five millennia.

As is the case with any work, we could not have done this alone, and we would particularly like to thank the following individuals for help or feedback on various parts of this work. Vlad Atanasiu—author of *Ugraphia*, (Atanasiu, 2025)—gave us very helpful comments on a very early draft of the book proposal, as well as some helpful pointers to resources about Arabic and Swiss German diglossia/digraphia. Alexander Gutkin read a complete draft and gave us a lot of insightful feedback. Professor KP Rao, Manipal Institute of Technology, provided information about Kannada typewriters, and we also thank Suma Bhat for putting us in contact with Professor Rao. Raiomond Doctor, always a source of information about a wide range of eclectic topics, helped us with many resources for typing in Brahmic scripts. Some of the information we received via Doc were from Mrs. Supriya Deepak Kulkarni, Shri Datta Typing Classes, Chalisgaon. Cibu Johny helped us with some of the details about Malayalam typing. And Aaron Bell engaged us in a very fruitful discussion about early Hangul printing technology. Thanks also to Jacob Dahl for confirming that the second author got the correct glyphs for the Sumerian proverb quoted in the dedication. Finally we thank Lucy Roark for her artwork in several of the figures as well as the cover art.

Finally we would like to acknowledge Alexandru Ciolan, our editor at Springer, for his support throughout the process of preparing this book.

Portland, Oregon, USA   *Brian Roark*
Setagaya City, Tokyo, Japan   *Richard Sproat*
Setagaya City, Tokyo, Japan   *Su-Youn Yoon*
                                                                        July 2025

Contents

Chapter 1
Introduction

1.1 Prelude

This is a book about a fundamental, ubiquitous technology, some form of which has, by definition, been around since the earliest recorded history, and which many of us use everyday and take for granted: written language and the tools used to write it. It is a class of technology whose common-use scenarios have changed dramatically over recent decades, so that many of us write every day in circumstances and mediums we would not have written in even just 25 years ago: on the bus, on a mobile phone even (unfortunately) while walking. Writing was once the province of specialists (scribes), but literacy rates are now high worldwide and barriers to written communication have never been lower.

That does not mean, however, that barriers do not exist when it comes to effectively communicating via text. Beyond literacy, there are many circumstances that can impede the kind of fast and ever-present text creation that many of us enjoy, and we will dive into these circumstances in the course of this book. Those actually facing such barriers have no need for them to be spelled out for them: They deal with them every day. It is rather for those to whom writing (or typing, texting, whatever you want to call it) is like swimming to a fish—automatic, constant, taken as a given—that this book is largely written.

How and why does the standard technology that many of us have come to rely upon fail to meet the needs of other would-be writers? What are the assumptions about language and physical movements upon which current day text entry technology relies, and where are those assumptions invalid? Such questions are not unique to the current state of technology, but rather have come up repeatedly over many hundreds of years, with the advent of various mechanical writing technologies that predated computation. Such technologies worked very well for some writing systems and individuals with particular skills and abilities, less well for other writing systems or people.

© The Author(s), under exclusive license to Springer Nature Switzerland AG 2025
B. Roark et al., *Tools of the Scribe*,
https://doi.org/10.1007/978-3-032-00831-2_1

The same is true for today's technologies (for many of the same reasons), and we look at some of the similarities and differences between the technologies of various eras and how accessibility is impacted. Along the way we will document a bit of what is going on under the hood of this technology that we collectively are increasingly incorporating into every aspect of our lives, yet which too few understand well enough to reason about.

In just the past couple of years an important new technology has entered the realm, namely Generative AI. This has the potential to be a godsend for overcoming the kinds of barriers that we alluded to above. But it also explicitly adds a new dimension to the problem, namely agency: If I wrote my essay using ChatGPT, to what extent is it my work? Such implications regarding agency have always been there for those who use assistance in writing, and we will explicitly make connections between current uses of Generative AI and other scenarios where assistance is useful or even indispensable.

While this is ultimately a book about technology, it interacts with many fields-of-study, including cognitive psychology and neuroscience, linguistics, ergonomics, anthropology and speech and language pathology, just to name a few. Writing is, after all, the process of converting language created in the brain to some kind of representation in the physical world, and hence touches on topic areas pertaining to any part of that process. We do not aim to provide a comprehensive survey of the ways in which all of these fields intersect with the topic at hand; rather to highlight how the technology has changed over time to meet (and sometimes not meet) particular needs, and how in turn new technologies have changed—and will continue to change— writing. Thus, while not being a textbook in, say, cognitive neuroscience, we may touch upon germane topics in cognitive neuroscience to highlight interesting characteristics of writing technology.

For example, the issue of character amnesia in Chinese.

1.2 Character amnesia

In February 2007, a Chinese fish farmer, HLD, suffered a stroke causing a lesion in the rear portions of his left superior and middle temporal gyrus (Bi et al., 2009). This is a part of the brain just to the front of the sensory-motor strip that sits roughly in the middle of the front and back halves of the brain. Since the trauma did not affect any of the brain regions normally associated with speech and language, HLD's ability to speak and understand was unimpaired, as was his ability to read. He was, however, no longer able to write. When presented with a Chinese character that he was perfectly well able to read, and then asked to write it after a delay, his responses ranged from extremely defective to no response at all; see Figure 1.1. When asked to copy a character directly, he was substantially better, but still showed errors relative to unimpaired controls. When it came to copying simple line

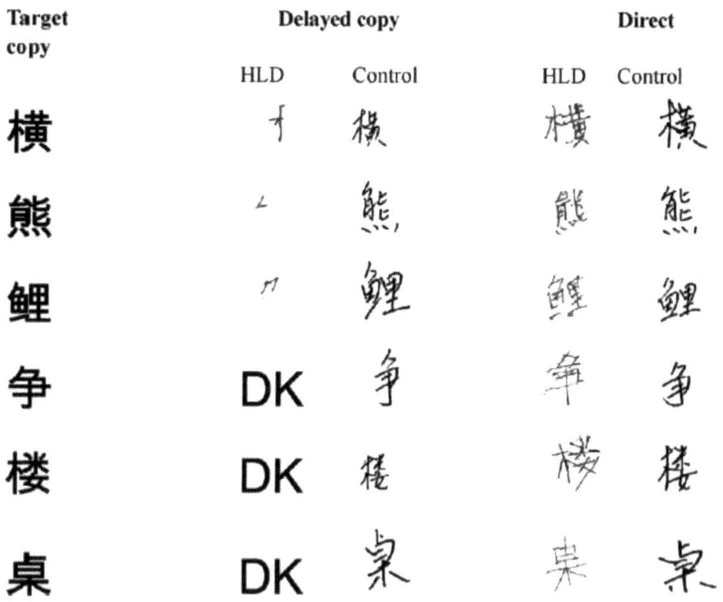

Target copy	Delayed copy		Direct	
	HLD	Control	HLD	Control
横	亅	橫	横	横
熊	㇐	熊	熊	熊
鲤	冂	鯉	鯉	鯉
争	DK	争	争	争
楼	DK	楼	楼	楼
桌	DK	桌	桌	桌

Fig. 1.1 A sample of HLD's responses to presented characters both when he was asked to reproduce them after a delay, and when asked to copy them directly. "DK" ("don't know") denotes no response. As can be seen, HLD's performance in the delayed copy situation was uniformly poor. In the case of direct copying, HLD did substantially better, but even here there were mistakes in several cases, an obvious case being the missing stroke at the top of 桌 'table'. In both tasks HLD's responses were compared to those of a control subject, who performed perfectly. Characters shown, from top to bottom, are: 横 *héng* 'horizontal', 熊 *xióng* 'bear', 鲤 *lǐ* 'carp', 争 *zhēng* 'conflict', 楼 *lóu* 'floor (story)', 桌 *zhuō* 'table'. Perhaps not coincidentally given HLD's profession in fish farming, the only character he was able to copy perfectly was 鲤 'carp'. Source: Bi et al. (2009), Figure 2, page 1196, used with permission.

drawings, for example a line drawing of a watch, HLD was slightly wobbly in his rendition, but was able to reproduce a recognizable rendition, even when drawing from memory. Only his knowledge of how to write Chinese characters appears to have been seriously impaired by his injury.

The location of HLD's injury, just to the front of the part of the brain most associated with motor memory, accords with other studies that have pointed to the same conclusion: Knowledge of how to write the complex characters of the Chinese script seems to be encoded as a set of motor instructions that mirror the way in which Chinese characters are taught. There are a set of prescribed rules for how characters are written and there is very little deviation from these rules. When one learns to write Chinese one starts by learning the eight basic atomic strokes, and then proceeds to learn how to

里 里 里 里 里 里 里 里

Fig. 1.2 Prescribed order of strokes for writing the character 里 *lǐ*. Source: `https://commons.wikimedia.org/wiki/File:%E9%87%8C-bw.png` GNU Free Documentation License 1.2.

combine them into more complex glyphs. Figure 1.2 presents the seven strokes used to write 里.

Knowing the strokes and their order is essential not only for writing characters correctly, but is also used as a fundamental organizing principle for the character lexicon, hence is required to find them in a dictionary. Traditional Chinese dictionaries are arranged according to the 'semantic' radical. The character 鲤 'carp' (shown in the third example of Figure 1.1) consists of two pieces, the left part 鱼, which on its own means 'fish' and here indicates that 鲤 is the name of a kind of fish; and 里, which on its own means a *li* (a traditional Chinese unit of distance), but here is used as an indicator of the pronunciation of 鲤, which is also pronounced *li* (or more specifically *lǐ* with a low tone). In a traditional dictionary, 鲤 would be grouped under the 鱼 semantic radical (traditional form 魚). The radicals themselves are arranged by stroke count, with simpler radicals provided first. A simpler radical than 鱼/魚 is 木, the 'wood, tree' radical, which is the semantic component of the first, fifth and sixth examples in Figure 1.1. And within each radical set, characters are arranged by the remaining portion of the character (the 'phonetic' component), again according to stroke count. In order to find 里 in a dictionary it is important to know that it is written with seven strokes, and in order to know *that*, it is important to remember how it is written. Those of us who have studied Chinese and have used a traditional Chinese dictionary have experience writing a character component in the air with our index finger in order to count how many strokes it has.

All handwriting of course requires motor memory. But what distinguishes writing Chinese from writing, say, English, is the amount of memory resources that must be devoted to encoding and maintaining the former system. The most efficient way to do that seems to be in terms of the motor actions required to write the characters, and these in turn have served as an indexing method for the system for the past couple of thousand years. The importance of motor memory in maintaining such a complex system implies that, as with any skilled set of motor actions, one can lose the ability not only by trauma, as in HLD's case, but also by simple lack of practice. And indeed this is the case: *Character amnesia*, as it has come to be known, is a growing problem among people who are native speakers of Chinese, who speak and read and write the language all the time, but crucially do not write the characters by hand, using instead a phonetic input method, on their computers or phones (Mair,

2010; Huang et al., 2021). Like HLD, such people—and there are millions of them nowadays—have no problem reading, but if asked to write a character with a pen, will often have forgotten exactly how to write it.[1]

Their dilemma is a direct result of a compromise necessitated by a mismatch between what is technologically ideal and what is technologically feasible. For Chinese writing has always presented a bit of a problem for technology, once that technology moved beyond the brush or pen. The modern compromise involves phonetic input systems. All Chinese children are exposed to phonetic systems before they are taught characters. In Mainland China the system is the Roman-based *pinyin* (literally 'spelling sound'). In Taiwan it is *zhuyin fuhao* ('symbols for attending to the sound') a system that was invented a century ago specifically for children learning to read, and which is colloquially called *bopomofo* after the names of the first four symbols in the script. To take an example from above, the character for 'carp' 鯉 would be transcribed[2] ⟨li⟩ in *pinyin* (⟨lǐ⟩ if you include the tone mark), and ㄌ一ˇ (where ㄌ is ⟨l⟩, 一 is ⟨i⟩, and ˇ represents the tone) in *zhuyin fuhao*. These systems have a small number of symbols and are easy to fit on a keyboard, even the more compact keyboard of a smartphone or wearable device. Conversion between a phonetic transcription system and characters is rarely unambiguous, but computational language modeling algorithms (about which we will talk more in later chapters) have become sufficiently good that it is usually possible for the system to guess given enough context what the intended characters are. And in any case the user is always presented with a choice of possibilities in case the input system's first guess is wrong.

There are certainly other ways to input characters. There are systems like *Cangjie* (named for the mythical inventor of Chinese characters) which are based on structural properties of the character, and which allow one to specify a character exactly. But they require learning, and few people bother to take the trouble to do so. And, within the last few years, handwriting recognition has matured so that one can use one's finger or a stylus on a device's touch screen to input characters by writing the old-fashioned way. But input using handwriting can still be a bit awkward, and for whatever reason does not seem to have had much penetration with users. Instead, given the universality of education in phonetic systems, and advances in natural language processing, phonetic input systems are the method of choice for nearly everyone who types in Chinese. But this choice of a technologically comfortable solution comes with a cost: Once people cease regularly using a pen, and start typing phonetically, the complex set of motor memories

---

[1] For example, the study by Huang et al. (2021) reports character amnesia occurring about 6% of the time for university students, and affecting 42% of characters.

[2] Throughout this book, where it is important to make the distinction, we use the linguistic convention of enclosing in angle brackets a string that is intended to be interpreted as the sequence of written symbols. Linguistic examples will be presented in italics. Thus *cheese* is the word "cheese", whereas ⟨cheese⟩ is the sequence of characters c, h, e, e, s, e.

that they learned in elementary school starts to erode. They become like HLD, but not as a result of trauma, just lack of exercise. They have gained ease of input with modern technology at the cost of losing practice with the neuromotor infrastructure needed to maintain their complex script. A true Faustian bargain.

1.3 The process of writing

As stated earlier, the topic of this book is technology underlying the process of writing, the process by which one converts a mental expression of a sentence to graphical symbols represented on a piece of paper, a whiteboard, or a computer screen. I want to write the sentence *My dog is very fond of avocados*, i.e., I want to translate it to a sequence of glyphs on a page or screen: ⟨My dog is very fond of avocados⟩. How do I do that, given my own abilities, and the technology at my disposal? This book is not about the psychological processes involved in producing writing (though we will mention results from psycholinguistics where relevant), but rather about how (1) the properties of the script, (2) the technology used to produce the text—what Mullaney (2017) terms *technolinguistics*, and (3) the abilities or disabilities of users, all interact to make the process of writing harder or easier.

Writing, whether one is typing on a QWERTY keyboard, or writing with a pen, is a complex learned skill. The process of writing can be made easier or harder depending on many things.

First there is the script, which can be more or less difficult to learn, remember or produce. Or the script itself might be easy, but the language's *writing system* might have complex spelling rules, like those of English, that have to be obeyed.

Then there is the technology of writing itself. The pen is an instrument that is largely indifferent when it comes to the script one uses, in that one can as easily use a pen to write Chinese as one can English.[3] But a typewriter, or a Linotype machine, discriminates. Popular images of gargantuan Chinese keyboards with thousands of keys were nothing more than caricatures (Mullaney, 2017), but real Chinese typewriters still required much more training and skill than typewriters for English or other alphabetic writing systems required. Computer technology has largely solved the problem by allowing one to use a standard keyboard that was originally designed for English to input Chinese phonetically. But this has the drawback discussed above: Chinese

---

[3] That said, there are certainly implications of using a pen, say, compared with more traditional implements. In the case of Chinese, one cannot render traditional calligraphy with a ballpoint pen. For that you need a traditional writing brush. But if one merely wants to communicate in writing using a particular script, then a pen can be said to be reasonably indifferent as to what that script is.

speakers who regularly use computers or phones to input text, forget how to write characters by hand.

And writing is more difficult for some populations than others. A person with severe motor impairment, such as that caused by cerebral palsy, may be unable to coordinate their movements well enough to type on a standard keyboard, or to produce speech intelligible enough to use an automatic dictation system. Many such people find it easier to use input systems based in MinSpeak® or other specially designed systems that allow the user to access words and phrases from (largely) semantically keyed dynamic keyboards. Some users, such as those with severe spinal injury, may not even be able to use those keyboards, and may only be able to control a device using a single button press, requiring them to answer multiple yes/no questions before arriving at the character they seek. How does one design a system that maximizes the speed with which such users can input text? The field of Augmentative and Alternative Communication (AAC) has devoted a large part of its focus over the last few decades to technological solutions to these sorts of problems. This has become a particular issue recently with the advent of Large Language Models, which can generate whole texts for you with a minimal initial prompt. The text may not always be what you would have written, but can often be close enough. However, this does bring up the issue of *agency*—an issue that actually has been of importance in the AAC field for decades, due to a debunked method known as *facilitated communication*. In this approach – described more thoroughly at the end of Chapter 6 – facilitators physically assist individuals in pointing to letters to produce text, the result of which controlled experimentation has shown to be entirely (if unconsciously) due to the facilitator rather than the individual. Accessibility is ill served if agency is not supported. We will touch upon such issues at various points throughout this book.

And what, again, about the complexity of the script? Most AAC systems have been designed with English in mind. How would a Chinese speaker with cerebral palsy input text? How about a Chinese speaker who can only use a single button press?

All of these questions come down to the same fundamental question: How do properties of the writing system and its written symbols interact with the technology that is used to deliver the symbols to the writing surface, and the abilities or disabilities of the person who is doing the writing? As we noted in the discussion of Chinese typing above, given the properties of the script, there is often a balance between what is technically feasible and what is ideally desirable.

1.4 Some technological advances

One does not need to look to the most challenging scripts or writing systems to see dramatic recent changes in the writing process, largely due to changes in technology. Forty-five years ago, written communication between New York and London did not have to rely upon delivery of a piece of paper across the water as it had 150 years earlier. However, in 1980, methods for the electronic transmission of text documents required that they either be input directly into some kind of specialized equipment (e.g., a teletype or perhaps a node on a pre-internet network), or scanned and transmitted via fax to a printer at the other end. Even with such options, somebody from New York on vacation in London would most likely prevail upon a physical medium (a postcard or maybe one of those thin blue sheets of *par avion* letter paper that folded into their own envelopes) to write home. Further, faxing required a document to scan, so in nearly all scenarios one would produce text on paper, usually either by handwriting or typing. Those of us old enough to remember writing while traveling in those days may have had some or all of the following experiences repeatedly: Traveling with a portable typewriter as an additional piece of luggage; writing with a pen until one's hand hurt and/or developing calluses on one's fingers where the pen rested; or spilling coffee on a nearly completed letter and just sending it anyway.

Twenty years later, at the turn of the century, one had a few more options at one's disposal in such a scenario, including sending emails—provided one had a laptop and internet service was available; or one could find a connected desktop computer to log into, say in an internet cafe or hotel. That email would typically be composed on a physical keyboard with constraints on the size of the symbol set and available font encodings, which would have been fine in English from London to New York—unless one was naïve enough to expect to be able to use a diæresis diacritic (or an ash ligature). Access at an internet cafe was most often timed and costly, so like the *par avion* letters from twenty years earlier, this was generally an asynchronous interaction, whereby one would compose and type an email and send it off, then return to look for a response at some later time.

Contrast this with how one might communicate via text from London to New York today. Perhaps one is standing on a crowded train on the Piccadilly Line in the evening, grasping a bar above one's head for balance with one hand and holding a mobile device in the other hand. With the thumb of the hand holding the phone, one inputs a relatively brief text message—a query or the start of an anecdote, less than 20 words. With a 5-hour time difference, it's mid-afternoon coffee break time in New York, so the recipient of the message may be available to respond nearly immediately. One can then have a brief, nearly-synchronous dialog, one person moving around in the London Underground, the other seated at a table in a cafe in New York. Such a dialog between familiars may be very informal, filled with slang, abbreviations, deliberately idiosyncratic spellings and non-linguistic symbol sequences such as ;-).

This is quite a different interaction from the trading of written or typed letters of 45 years ago (Crystal, 2006). Texting with one's thumb—perhaps with the off hand due to the constraints of sharing space on a crowded train—is hard. The virtual on-screen keyboard is small and the demands on fine motor control are heightened relative to typing on a physical keyboard, resulting in more frequent mistakes that must be corrected prior to sending the message. Further, there is typically software running to detect and autocorrect common errors, which may itself introduce errors into the message. One must attend to the screen of the device in ways one doesn't when touch typing on a laptop. Predictions or completions are often supplied by the keyboard and selected or not; with Generative AI, predictions or completions may be as lengthy as one desires, but may not always be what one wants, and one is faced with the decision of how much of the output to accept as *one's own*. The difficulty inherent in the writing scenario is deemed worth it because of the interaction. The resulting dialog is very different from exchanging letters, and so is the process of writing, though some important similarities persist, such as the layout of the keys on the keyboard.

The point of this brief segue into modes of personal written communication in English over the past decades is this: Even for those writing in languages well supported by writing technologies over the years, it is clear that, depending on the scenario, effective writing can be difficult. Writing under time pressure with a keyboard that is too small to make all the required symbols immediately accessible makes message creation complex. But writing has always been a complex task, something that may not occur to someone who typically writes under relatively ideal circumstances: at leisure with typical motor control and the writing/typing/texting expertise to produce language with ease; and in languages well supported by writing technology. One does not need to make this case quite so explicitly to those writing in Arabic, Hindi or Japanese, for whom writing technology has been generally less of a perfect fit. Even so, today speakers of these languages may also be standing on the Piccadilly line texting with their friends and families in their native languages and scripts, each with their own challenges.

People all around the world now carry out nearly synchronous conversations using text, sometimes person to person (texting), sometimes in a small group (various chat apps) and sometimes in public (social media). This puts a premium on fast writing, something that is easier in some writing systems than others and for some individuals than for others. Fast production of text, however, is not a new problem, and has its correlates in typesetting, stenography and assistive technologies, all of which were scenarios in which speed of composition was an important goal. Some of these areas of technological innovation were hugely successful in the West, but were less successful in other parts of the world, such as Asia, where differences in writing systems made simple solutions to fast text production elusive. Many of these same problems are at play in today's text-centric world. However, the existence of gargantuan amounts of electronic text permits the use of natural language technology that in turn can help make text production faster and/or more accurate.

In this book, we aim to present writing technology past and present with a broad focus, discussing common technology but also technology serving communities of writers that were (and sometimes still are) not otherwise well served. For example, text is the principal communication modality for many with severe motor disabilities: How does one type if one cannot easily or reliably point to a specific key on a keyboard? Alternative input methods can be devised to permit typing in such circumstances. What if Jack Kerouac had been a Kannada speaker?[4] Might Truman Capote's famous quip about writing versus typing[5] still have been true? Probably not, since given the nature of its script (an *abugida*, or *alphasyllabary*), Kannada is more difficult to type in than English, and typewriters were thus not nearly as ubiquitous as they were for English and other languages using alphabetic writing systems. How about now, if Jack were texting in Kannada from a mobile device on the Piccadilly Line? Yes, texting is as common for South Asian languages as for other languages, though to type the word for *hello* in Kannada ⟨ನಮಸ್ತೆ⟩ one might end up using a QWERTY keyboard on the mobile phone to type the word ⟨namaste⟩ and have it automatically converted to the Kannada script. Thus, as in Chinese, alternative input methods have been devised, though typical solutions in South Asian languages differ in interesting ways from those in Chinese. Details on such languages and scripts—what makes them difficult to type, what systems exist to realize the languages in the Latin script—will be covered later in the book.

There are many considerations that determine the complexity and difficulty of writing, including the complexity of the writing system and script, the complexity of the sequence of actions required to write or otherwise produce the script, and the requirements of the specific writing scenario.

We present examples of these points in the next two sections.

1.5 Technology and script complexity

A simple illustration of how a given technology can interact with a given set of design choices for a script is found by looking at a brief history of typing in Hangul ('great writing'), the phonemic writing system for Korean.

---

[4] Kannada is a Dravidian language spoken in southern India.

[5] Initially speaking of writers without style generally, in an interview published in the Paris Review (Hill, 1957), Capote said, "But yes, there is such an animal as a nonstylist. Only they're not writers. They're typists." This was later variously paraphrased – by Capote himself and others – and applied to Beat Generation writers as a group and, eventually, also to Jack Kerouac specifically.

Hangul was invented under the guidance of King Sejong the Great (李祹 *i-do*, 1397-1450), the fourth monarch of the Joseon dynasty, and promulgated in 1446 in the 訓民正音 (*Hunmin Jeong'eum*) 'Proper Sounds for the Education of the People'. By replacing the Chinese writing that had been used for centuries in Korea with a phonetic system based on 28 letters, Sejong hoped to spread literacy to the common people (whom Sejong termed the 故愚民 *go u min* or 'uneducated folk'), most of whom had no opportunity to learn the Chinese language and writing system. While the scholarly class largely scorned Hangul—why would they give up a hard-won privilege based on years of inculcation in the Chinese classics (see also Chapter 7), its use spread among the people. In the early 20th century, it became the sole writing system used for Korean, and though Chinese characters were still taught, their use declined. Today one sees very few Chinese characters in Korean text, except the occasional lone character used for largely stylistic reasons in newspapers,[6] or as a gloss for a Sino-Korean word (a word of Chinese origin) in, for example, a Wikipedia article.

Since there are scarcely more Hangul symbols than there are letters in the English alphabet, the system is easy to learn and, as one can imagine, easy to adapt to a keyboard. But there is one glitch: in Sejong's system, the symbols were never intended to be written linearly like letters in English, but rather arranged into syllable-sized blocks. There was a simple reason for this: the traditional writing system used in Korea was Chinese, and each Chinese character represents a syllable. So it seemed natural to group the basic symbols of the alphabet into syllables.[7] And while Hangul looks quite different from Chinese, if one glances at a page of Hangul text, each syllable occupying the same amount of space as the next, it looks superficially like a page of Chinese text. No doubt these considerations were aimed at making the system look less "foreign" to the people who were already literate.

But the result of this decision is that if one wants to write the word *manhda* 'many' one does not write ㅁ ㅏ ㄴ ㅎ ㄷ ㅏ , that is ⟨manhda⟩, but rather 많다, comprising six Hangul letters (*jamo*) grouped into two syllable glyphs: ㅁ (upper left), ㅏ (upper right), ㄴ (lower left), and ㅎ (lower right) combine to form 많 ⟨manh⟩, and ㄷ and ㅏ together form 다 ⟨da⟩. If one is writing with a pen, of course, this makes minimal difference: one just has to understand the rules by which one combines the letters into syllable blocks. And if one jumps from the pen to the modern computer or hand-held device, there is also no problem: one can type the individual letters on a regular keyboard using any of a few different layouts, and algorithms combine those

---

[6] For example the headline 美 캘리포니아 산사태 사망자 19명으로 늘어 …5명 실종 (**Donga.com**, January 14, 2018) 'United States, California landslide death toll climbs to 19, five missing', starts with the single character 美 *mi*, shorthand for *migug* (美國, Hangul: 미국) 'America'. The remaining text of the headline is in Hangul, and Hindu-Arabic numerals.

[7] Hangul may well have been influenced in its design by the Brahmic scripts of India, which also arrange their basic symbols into syllable-sized chunks, so this may have been another reason for the choice.

letters into syllables, and display them using one of the more than 11,000 theoretically possible Hangul composite syllables supported in Unicode, with an aesthetically pleasing result, assuming the typeface designer has done their job properly.

The problem came with a technology intermediate between the pen and the computer—the typewriter. When one depresses a key on an English mechanical typewriter, the strike plate hits the ribbon, impressing the letter onto the paper. The machine then automatically advances to the next position. This mechanical procedure typically leads to monospaced text, i.e., it types `cheese` since each letter is allocated a fixed width. While this has a distinctive look that one may like or dislike, it leads to legible text. However, such a mode of operation would not work for Korean, since if one wants to type the syllable 군 ⟨gun⟩ for example, the carriage should not advance between the ㄱ and the ㅜ or between the ㅜ and the following ㄴ. Similarly for 긴 ⟨gin⟩ one wants the ㅣ to appear to the right of the ㄱ, and the ㄴ to appear underneath the two previous letters. Plus if one wants to get really fancy, the reader will have noticed that the form of the ㄱ ⟨g⟩ is slightly different in 군 and 긴: there are two basic *allographs*[8] of ㄱ, that depend on whether the following vowel is a vertical vowel written to the right of the ㄱ, or a horizontal vowel written underneath. More complex syllables such as 끓 ⟨kkeulh⟩ present further issues.

With the advent of the typewriter, there was even talk of "modernizing" Korean orthography by linearizing it, thus removing these problems. But as with similar attempts to "modernize" other scripts, such as proposals for Chinese that also coincided with the introduction of typing technology, as discussed in Mullaney (2017), these never gained traction.

To get a functional Korean typewriter, one at least needed to divide the set of jamo into:

- initial consonants, which always appear at the top or left depending on the following vowel;
- final consonants, including consonant pairs like the ㄹ ㅎ in syllables like 끓, which always appear at the bottom;
- and vowels which, depending on the vowel, must be either to the right of, or below the consonant they logically follow.

While striking the keys for the jamo making up a syllable, the carriage of the machine must effectively remain stationary. The glyphs on the bars had to be carefully designed so that the characters would land in the right places so that if one were typing 긴 ⟨gin⟩ the ㅣ ⟨i⟩ would appear to the right of the ㄱ ⟨k⟩ but if one were typing 군 ⟨gun⟩, the ㅜ ⟨u⟩ would appear underneath. Once one was done composing the complex glyph one could proceed to the next glyph by hitting the spacebar.

---

[8] Allographs are variants of the same symbol or grapheme, such as lowercase ⟨a⟩ and uppercase ⟨A⟩ in English.

While this worked, the result did not always look very pretty.[9] An example of a Korean typewritten document is the Korean Armistice Document, part of which is shown in Figure 1.3. Note for example how the final consonants in Kim Il-sung's name ⟨gim il-seong⟩, which appears typewritten as

roughly halfway down the page on the left, are skewed to the left relative to the rest of the syllable, compared to the more balanced appearance in a modern computer typeface:

김일성

The same problems found with Korean typing can be found for other scripts, such as the Brahmi-derived scripts of India and Southeast Asia, e.g. the Kannada script mentioned earlier. For all of these scripts the size of the alphabet is small enough that it is practical to type with a reasonably-sized keyboard (though as we shall see in a later chapter, typing on a phone can still sometimes be a challenge). But the symbols combine in complex ways, so designing a typewriter was much harder than designing one for English or, say, Russian.

1.6 Expertise, technology and gesture complexity

Jean-Dominique Bauby's 1997 memoir *The Diving Bell and the Butterfly* (Bauby, 1997) is a beautifully written account of the author's experiences following a massive stroke that left him "locked-in", i.e., unable to move or speak. Writing a book while in such a state of physical incapacitation is no mean feat; how did he do it? He used his one remaining reliable motion—the blink of his left eye—to dictate the memoir to a communication partner. Those of you familiar with this powerful book (or the acclaimed film of the same name) will know their method of eyeblink communication. Those who are not familiar with this might try to guess how this worked, and might end up guessing alternative methods that were in fact not used by Bauby and his communication partner, such as Morse code.

While it was not used by these individuals, Morse code is sometimes used as an assistive technology. For the purposes of this preliminary discussion, suffice to say that in Morse code each letter in the Latin alphabet is assigned a code, consisting of a sequence of dots and dashes. For example, the letter ⟨t⟩ is a dash, the letter ⟨e⟩ is a dot, and the letter ⟨n⟩ is a dash followed by a dot,

---

[9] Mullaney (2017) discusses similar problems in Chinese typewriter designs that attempted to break complex characters into their component parts. The technology sort of worked, but the result did not look very aesthetically pleasing.

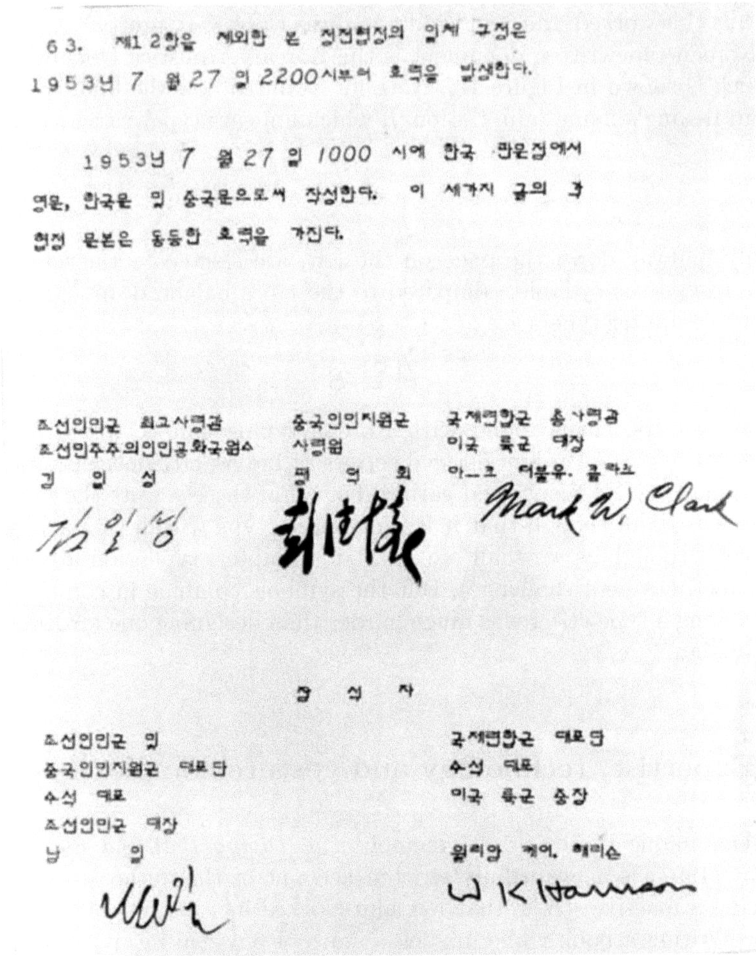

63. 제1 2항을 제외한 본 정전협정의 일체 규정은
1953년 7월 27일 2200시부터 효력을 발생한다.

　　1953년 7월 27일 1000 시에 한국 판문점에서
영문, 한국문 및 중국문으로써 작성한다. 이 세가지 글의 구
협정 문본은 동등한 효력을 가진다.

조선인민군 최고사령관　　중국인민지원군　국제련합군 총사령관
조선민주주의인민공화국원수　사령원　　　　미국 륙군 대장
　김 일 성　　　　　　　펑 덕 회　　　아―크 더불유. 클라크

Mark W. Clark

참 석 자

조선인민군 및　　　　　국제련합군 대표단
중국인민지원군 대표단　　수석 대표
수석 대표　　　　　　　미국 륙군 중장
조선인민군 대장　　　　윌리암 케이. 해리슨
　남 일

W K Harrison

Fig. 1.3 Part of the Korean Armistice Document, an example of typewritten Korean. Source: https://commons.wikimedia.org/wiki/File:Korean_Armistice_Ko-Text_1953.jpg. Image is in the public domain.

so the word *ten* would be transmitted: ■ ● ■●.[10] This sort of code is suitable for communication with eye blinks. In fact, it was famously used with eye blinks in a very different scenario than that faced by Jean-Dominique Bauby. In 1966, American prisoner-of-war (and future U.S. senator) Jeremiah Denton, Jr., was interviewed as part of a North Vietnamese propaganda broadcast shown on American television. While he responded verbally to posed ques-

---

[10] Typically dash and dot are differentiated by duration, with a dash lasting three times as long as a dot. Between each letter a pause of duration equal to a dash indicates a letter boundary, and between words a longer pause.

tions, he was also blinking a message in Morse code, spelling out the word *torture*.[11] However, the code must be learned by both the individual and the communication partner, and thus may not be pursued if other, easier to master methods are available.

Another possible reason for not using Morse code requires comparing it with the method that Bauby and his communication partner actually used which is known as partner-assisted scanning (Garrett et al., 2007). In partner-assisted scanning, the communication partner presents (visually, auditorily or both) each option (in this case each letter in the alphabet) to the communicating individual, who indicates (in this case via eye blink) when the intended option is presented. Thus, with the Latin alphabet, one may have to present all 26 letters before the intended one is found. While this worst case seems very inefficient, on average the number of items that must be presented is much lower, particularly if the letters are presented (as Bauby's partner did) in descending frequency order, i.e., the most common letters first. The overall efficiency of such an approach, of course, also depends on the speed of presentation, which in turn depends on what is comfortable for the individual.[12] So why would this be preferable to Morse code for somebody with locked-in syndrome, even if that person knew Morse code? Note that with partner-assisted scanning as we have described it here, each letter requires exactly one blink, but with Morse code, most letters (all but two) require multiple blinks (of varying duration), corresponding to the specified code for that letter. For somebody dealing with a severe motor disability such as locked-in syndrome, each blink requires concentration and effort to the extent that, frankly, either method can be found to be exhausting. Even had Bauby and his communication partner known Morse code, it would have entailed significantly more blinks for the same message, which may have made the method infeasible for him, even if in theory the time to construct the message would be less.

This brief discussion of one person's need for alternative text generating methods brings out several points that should be emphasized, as they will become key themes throughout the book. First, while the above-presented method was partner-assisted, computer-assisted scanning as a general method of text entry is also very common in AAC. In such methods, there is an elicitation event (e.g., something is presented), and the response is interpreted relative to what was presented. Presenting one letter at a time for a yes/no response is a particularly simple setup, and other, more complicated, setups are quite common, such as highlighting rows, columns or other contiguous

---

[11]   https://www.nytimes.com/2014/03/29/us/politics/jeremiah-a-denton-jr-war-hero-and-senator-dies-at-89.html

[12] Stephen Hawking used a computer-assisted scanning method for his text entry, and had apparently developed the expertise to set the scan rate to be very fast, i.e., it was comfortable for him to deal with items being presented for a very short amount of time before moving on. His system will be touched on in a discussion of scanning systems generally in Chapter 5.

regions of a keyboard or grid of symbols, and recording the response ('yes' meaning that the target symbol is in the highlighted region). One can see it as a sort of 20-questions dialog between the individual and the partner or device eliciting the text input. Both the form of the elicitation and the type of expected response can be tailored to the specific capabilities of the individual. Such methods will be presented in detail in later chapters, and ultimately we would like to situate computer-assisted text entry devices of this sort in relation to other writing technology.

The role of expertise in producing text is also an element of the above discussion that should be emphasized. The base expertise required is literacy, and this is an expertise that a higher percentage of the world's population has now than ever before. According to a recent UNESCO report on literacy, the worldwide adult literacy rate has risen over 30% between 1950 and 2015, from 55.7 to 86.2 percent (UNESCO, 2017). Expertise in producing text, however, goes beyond basic literacy to include such skills as touch typing on a physical keyboard, thumb typing on a virtual keyboard or (less commonly) Morse code. Learning a new expertise—e.g., learning to be proficient in Morse code, or switching from a standard QWERTY keyboard layout to a different layout—can yield benefits in speed, accuracy, or comfort, but generally at a cost and usually with some tradeoffs. Technology has greatly reduced the level of expertise required for communicating by text, all but eliminating the need for the skilled text producing professions of the past, from scribes and scriveners writing by hand, to typists and typesetters producing documents via more-or-less complex machinery. Still, specific writing scenarios, such as writing with limited motor control, may require new techniques and new areas of expertise.

While we presented scanning as an alternative to Morse code, note that the scanning method itself defines an implicit code for each letter in the alphabet. If each letter in the alphabet is being presented in a particular order for the individual to reply 'yes' or 'no' (either through blinking or some other means), then the 'code' for each letter can be seen as the sequence of yes/no responses needed to select that letter. If we make 'yes' 1 and 'no' 0, this corresponds to a binary code for each letter in the alphabet. In general, the shorter the code, the faster and/or easier the text input process will be, so we will see many methods for reducing the ultimate *code length* for text entry.

A final point that should be emphasized here and will be returned to throughout the book is the role of prediction in writing technology. When discussing his mode of scanning, Bauby writes:

> Meticulous people never go wrong: they scrupulously note down each letter and never seek to pierce the mystery of a sentence before it is complete. Nor would they dare dream of finishing a single word for you. Unwilling to chance the smallest error, they will never take it upon themselves to provide the 'room' that follows 'mush,' the 'ic' that follows 'atom', or the 'nable' without which neither 'intermi' nor 'abomi' can exist. (Bauby, 1997, pp. 29-30)

Indeed, human communication partners can provide quite useful and accurate predictions about the word, phrase or even sentence that is intended, and such a prediction is quite well suited to the sort of binary yes/no question that forms the basis of this sort of scanning. Machines can also make predictions, and indeed the predictions available from Generative AI can be quite stunning. So a similar sort of human/machine dialog can occur as can occur between two people, also with the potential for substantially faster communication. Prediction, however, also plays a role in more quotidian text interactions, such as thumb typing on a small virtual keyboard. The precision of the points where the keys are actually touched (touchpoints) can vary quite widely, and such an interface, in contrast to physical keyboards, must generally assign a probability that a key was intended, instead of simply registering whether it was touched or not. Similarly, for the phonetic input systems for Chinese mentioned earlier in the chapter, such as pinyin, the intended character for a given input is generally uncertain, but statistical models can predict the likely intended sequence. We will return to the kinds of predictions that must be made in these and other scenarios throughout the book, as well as some accessible discussions of how such predictive models are estimated and used in practice.

This brief discussion of human and/or machine prediction as the means to faster text entry also highlights another important issue in text entry: privacy. Imagine sending a private text message with the help of a communication partner, who speeds up the entry by guessing what you are going to say next. If that person was able to help you write your message in 30 seconds rather than the 15 minutes it might take you to input the message without their assistance, then you might very well accept their help. If a human/machine dialog of the sort we described above could provide nearly the predictive utility as the human communication partner, then that might be a way to speed up typing without having to share the details of your text message with another person. Even with computer models, however, there is a risk to privacy if care is not taken, since data is required to train models and the system making predictions must access a message while it is being constructed. Training and using models in ways that preserve privacy is an important aspect of writing these days, which we will return to later in the book.

1.7 The rest of this book: Some preliminaries

We wrap up this chapter by touching on a few points that relate to the general themes of the book, and which will be addressed in more depth in the ensuing discussion.

1.7.1 Writing is not language

Though the two are often confused, writing and language are distinct. Probably the most oft-quoted statement about this distinction is due to Leonard Bloomfield who, in his book *Language* (Bloomfield, 1933), stated that "[w]riting is not language, but merely a way of recording language by visible marks."

In other words, writing is merely a linguistic technology, a way of representing language, but not language itself. A musical score is not itself music, merely the way of representing music that can be interpreted by the musician as music; and in a similar way, writing is a representation that can be interpreted linguistically by someone who is literate in the language in question. Let us take a simple example to make the point. For centuries the Turkish language had been written with an adaptation of the Arabic script, but this changed virtually overnight in 1928 when, as part of a broader initiative of Westernization, Mustafa Kemal (Atatürk) promoted a Romanized version of Turkish writing, which subsequently became the sole medium for writing Turkish. Obviously the Turkish language itself did not change overnight, though the introduction of Roman as opposed to Arabic script, coupled with the intentional Western orientation, did serve to distance Turkey from the Arabic sphere of influence, meaning that loanwords would henceforth come more easily from Western languages than (as they historically had) from Arabic.

1.7.2 Types of writing systems

Writing systems differ a lot in how they represent language, but they all have one thing in common, namely that they all have some mechanism for representing sound. We started our discussion in this chapter by looking at how the complexity of Chinese writing can affect the memory of how to write Chinese characters, and we noted there how most Chinese characters are composed of two components, one that represents something about the meaning, and the other that represents the pronunciation. The representation of pronunciation is often not very good, especially in Modern Chinese languages, and Chinese characters are a bit like a graphical game of charades: 鲤 *lǐ* 'carp' is a kind of fish (鱼), and sounds like *lǐ* (里). 熱 *rè* 'hot' has something to do with fire (灬) and sounds a bit like *yì* (執). But lousy though the phonetic representation is, nonetheless it is a critical factor in the way Chinese writing works: fully 95% of the thousands of characters that have ever been invented operate on this semantic-plus-phonetic principle. The situation was identical in the other original separately invented writing systems: Mesopotamian, Egyptian and those of Mesoamerica. In each case besides the symbols that represented something about the meaning of the words being represented, there were also symbols that represented sound.

This universality of phonetic representation is what writing systems scholar John DeFrancis (1989) termed the "diverse oneness of writing systems", and it gives us a foothold into classifying writing systems and understanding some of the sources of their complexity for the writer.

Thus, in Chapter 2, we will explain how writing represents linguistic and in particular phonological information. One aspect of this representation involves how the basic symbols of the script are arranged on the page which, as we will see in Chapter 3, can have a profound effect on how easy or difficult it is to write those symbols, depending on the technology one is using. And in Chapter 4 we delve deeper into how simple or complex the representation of linguistic information is in a writing system: In some writing systems this is very straightforward, in others much less so. While this may not affect the mechanics of say, typing, it does have implications for language technology. In English, or Japanese, there are many homographs (words that sound the same but are spelled differently), and people often make mistakes. This means that a "smart" input system must be on the lookout for a user's spelling errors.

1.7.3 Writing technology and the writer

Much of this book will focus on writing technology, which in its simplest characterization, involves two parts, namely the implement *with* which one writes, and the medium *upon* which one writes. Across both space and time, both of these have varied widely. Styluses of various types, pens, brushes, carved wood blocks, moveable type, and keyboards connected to typewriters, Linotype machines and computers have been just some of the tools which people have used to construct the written word. Clay, stone, papyrus, bamboo strips, palm leaves, wooden tablets, wax tablets, cloth, paper and finally electronic memory with an auxiliary (usually visual) display, are just some of the kinds of mediums on which words have been written.

Each of these has its advantages and disadvantages. Inscriptions on stone have a permanence that has been important for upholding the authority of political systems throughout the ages, but obviously are not very practical as an everyday writing medium. Widespread literacy depended in large part on the invention of lightweight materials (papyrus, paper, bamboo strips and so forth), though this in itself was not enough. As we have noted, some writing tools such as pens are more "democratic" about what kind of writing system one uses than others, such as the typewriter. Some, such as modern computational interfaces, render the playing field more level: It is almost as easy to type Chinese on a standard keyboard as to type English. But as we noted above this comes at the cost of writers' forgetting how to write characters by hand that they are quite able to recognize. The interaction of the implement, the medium, the writing system and the writer, will be a central focus of this book.

But then there are the writers themselves. The earliest writing systems were associated with a professional technician class—scribes, and this class of technician has survived in various forms throughout the millennia, even as writing has become democratized: the scrivener or copyist, the telegrapher, the Linotype operator and finally the modern court stenographer are all versions of the ancient scribe, someone whose profession is writing and who has usually also mastered a technology that few others know how to use. For most people, writing is usually a slower process than speech, and scribes—and writers more generally—have developed various kinds of abbreviatory devices to speed the process. These may be actual abbreviations, such as the Tironian shorthand 7 symbol or various precursors of the modern "&" used in Medieval manuscripts to represent the common word *and*; or totally different systems such as Pitman shorthand or the short-lived Palm OS Graffiti, which reduce the number of strokes one needs to write a symbol.

Congenital disabilities or disabilities caused by physical trauma, as in the case of Bauby discussed above, also often conspire to force the writer to adopt novel approaches to writing. Ultimately the issue comes down to the fact that the writer wishes to communicate information via a written medium. Given the nature of the medium, the abilities of the writer, and the nature of the writing system that is being used, what is the most efficient way to effect the goal of communication? Especially when advanced technology or severe disabilities are involved, the approach can often be quite inventive—and in fact can often serve as a precursor to methods adopted by the general population, such as T9 input for ambiguous keyboards, as we will show in detail in the ensuing chapters. In later chapters of the book we begin to look at the complexity of the required gestures to input text, in manual, mechanical and computational settings, and at specific scenarios and constraints on writing, such as writing under space and/or time pressure or in a heavily multilingual setting. Thus in Chapters 5 and 6 we will discuss, respectively, gesture complexity, and the problem of writing under constraints imposed by limits on space and time.

Chapter 7 deals with the issue of writing in multiple languages—multilingualism, particularly in cases where the writer may speak more than one language, but is more comfortable writing in one—say an educated native speaker of Kannada, who is fully comfortable typing in English, but struggles to type in his or her native language.

Indeed the writer may not even be a native, or even particularly competent, speaker of the target language, which brings us to the topic of Chapter 8, where we discuss writing technologies for non-native writers. One of the themes there will be how recent AI technology is both incorporated into second language writing, and how in turn that technology is affecting second-language education.

Chapter 2
Writing, speech and graphical representation

2.1 Prelude

The topic of this book is the act of writing, how a message to be communicated is transferred from the writer's thoughts to a surface (be it paper or a computer screen), and the cognitive and technological impediments to that process. The technological details vary greatly across different conditions but one thing they have in common is the technology that underlies the whole act of writing: writing itself. To understand how printing, typing, computer input systems, or the writer's knowledge of their language and writing system influence the act of writing, it is therefore important to understand something about how writing itself works. This is the topic of this chapter.

Writing systems encode information. Essential to understanding how they work is understanding *what* information they encode and *how* they encode it. An obvious starting point is that written symbols encode something about the language they are used to write: writing is a *linguistic symbol system*. But we can be much more specific than that: for the most part, written symbols represent sound—phonology—though in various writing systems a large portion of the symbols may also represent other sorts of linguistic information—written symbols representing whole words, or even symbols that encode abstract meanings. Nonetheless, phonology is central, indeed critical in how writing systems evolved.

As for *how* the information is encoded, writing systems display a wide range of complexity in how straightforwardly linguistic information is represented, with some writing systems being very *shallow* and others much more *deep*, terms that we will elucidate below in Section 2.3.2. There is also a wide range of differences in how the basic symbols of a script are arranged on the page, with some scripts, such as the Roman script used in English, opting for a simple linear concatenation of symbols, with other scripts, such as the Brahmic scripts of India and Southeast Asia showing a much more complex arrangement of symbols, a topic we discuss in some depth in Section 2.3.3.

B. Roark et al., *Tools of the Scribe*,
https://doi.org/10.1007/978-3-032-00831-2_2

Again, these various facets of writing systems and scripts will be shown to be relevant for writing technology, as we will explore in more depth in subsequent chapters.

2.2 What defines writing

In this book we use the term 'writing' with a very specific meaning, namely *a means of conveying in graphical form what one might express in speech*.[1] This pretty much accords with Leonard Bloomfield's 1933 definition of writing as a "way of representing speech by means of visible marks." This definition is by no means uncontentious, and many scholars with an interest in writing systems differ on what exactly the term 'writing' means—see also DeFrancis (1989). Indeed some prefer a much more 'inclusive' definition that would include apparently *non-linguistic* systems such as *khipu* accounting symbology (Urton, 1998, 2010, 2017; Hyland, 2021).

One of the authors has discussed this controversy extensively elsewhere (Sproat, 2023), and we do not propose to revisit it here. We merely note two reasons why we adopt the narrow definition of writing that we do, the first empirical and the second practical.

The empirical point is that, no matter what one may feel about broader notions of writing, it remains true in every single instance of recognized independent inventions of writing that the key insight that was discovered was that one could use symbols to write the sounds of words rather than just their meanings. This point is not open for debate: it is a simple fact. The discovery was made in Mesopotamia; it was made, possibly independently in Egypt, apparently independently in China and certainly independently in Central America. In Mesopotamia, the one region where we have enough archaeological data for the early development of writing, it was clear that the earliest writing developed from an accounting system that had previously contained only ways to write concepts—numbers and commodities, but soon evolved into a system that allowed one to also write names, verbs, and eventually grammatical markers. This extension depended on the ability to write sound, since coming up with a system that could encode all of these categories using purely meaning-based representation would have been a daunting task.

In fact this principle applied not only to the few cases in history where writing was invented ex nihilo by a culture that had no contact with another culture that already had writing. It also applied in cases where the inventors were inspired by already existing writing systems, but who did not know how those systems worked. Thus when Sequoyah (ca. 1770–1843) invented

---

[1] The focus on speech here ignores *signed languages*. Written "phonetic" scripts do exist for signed languages, for example Stokoe's notation system for American Sign Language (Farnell, 1996), or Sutton notation (`https://en.wikipedia.org/wiki/SignWriting`), but these do not seem to be widely used in the signing community.

the writing system for Cherokee, he knew about English writing but did not understand how English writing represented the English language. In fact, at first he attempted a purely *logographic* system, where each word would have its own sign, but quickly gave up on this approach as impractical, inventing instead a syllabary.

So, if a large part of writing involves representing sound, then it is not at all an unreasonable characterization of writing to say that it is a "way of representing speech by means of visible marks." If a symbol represents a meaning directly—say a picture of a cow's head used in an accounting system to represent a cow as a commodity—then it is at least debatable whether this represents speech at all. If nothing else, a picture of a cow's head does not tie the symbol to any particular language. Indeed this is one reason why it is notoriously difficult to be sure of what language the users of the earliest prelinguistic accounting systems in Mesopotamia spoke. Once phonetic symbols were developed, on the other hand, it became clear that the scribes were intending to represent a particular language. To put it in modern terms, a cow emoji 🐮 can be 'read' in any language. The written form ⟨cow⟩, representing /ˈkaʊ/, can only mean that the writer intends English.

Our practical reason for adopting the narrow definition is that, for a large class of people with disabilities, writing is in fact a substitute for speech. This includes people with cerebral palsy, amyotrophic lateral sclerosis (ALS), or locked-in syndrome, who lack the ability to speak, and are often more generally limited in their motor abilities. For these people writing, often converted by text-to-speech technology into speech, is the *primary* means of expressive language. So while it may be more inclusive to include a broader range of symbol systems under the rubric of writing, for people with disabilities that prevent them from speaking, what is of interest is ways to represent in graphical form that which can be said in speech—exactly our narrow definition.

2.3 What writing encodes, and how it encodes it

This chapter does not aim to give a comprehensive introduction to writing systems. Rather we merely intend to highlight the features of scripts and writing features that are most germane to our main theme. The reader who seeks a more comprehensive treatment of the subject fortunately has at their disposal a number of book-length treatments of the topic. General introductions in English to writing systems from various points of view can be found in: Gelb (1952); Moorehouse (1953); Gelb (1963); Diringer (1958); Sampson (1985); Coulmas (1989); DeFrancis (1989); Drucker (1995); Coulmas (2003); Rogers (2005); Gnanadesikan (2009); Powell (2009); Sampson (2012); Daniels (2018); and Meletis & Dürscheid (2022).[2] A particularly good recent treat-

---

[2] An overview of the literature on writing systems through 2018 can be found in Gnanadesikan & Sproat (2018).

ment of Chinese writing and its derivatives is that of Handel (2019). Books
covering the early genesis of writing include Woods et al. (2010) and Glassner
(2000, 2003). The most systematic coverage of a wide range of scripts and
writing systems, both ancient and modern remains Daniels & Bright (1996).
Decipherment of ancient scripts is treated in Pope (1975, 1999); Parkinson
(1999); and Robinson (2009). Some more popularly oriented books on the
topic include Robinson (2007) and Ferrara (2022). Finally, a recent book by
the second author (Sproat, 2023), discusses the relationship between writing
systems and graphical symbol systems more generally.

For the present discussion, there are three aspects of writing systems that
are of particular interest:

- What linguistic information a writing system encodes.
- How 'transparently' that information is encoded.
- And how the basic symbols of the script are arranged.

Before we elaborate on these three points though, we need to be clear
about the meaning of two terms that we have used above, but have not
really defined: *script* and *writing system*. A script is a set of symbols, as
well as rules for combining those symbols into well-formed strings (the third
point on the list above). The Roman alphabet, for example, is a script. How
that script is *used* to represent a given language varies of course with the
language, and thus when we speak of a language's writing system, what we
are speaking about is a pairing of a script with a language, including the rules
to represent the latter in the former. The Roman alphabet has more or less
the same set of symbols whether it is used for English or Dutch, for example,
but the ways in which it is used in the two languages obviously differ, so that
one can talk of the English writing system or the Dutch writing system as
separate entities though they both share the same script.

2.3.1 What linguistic information do writing systems encode?

The first point on the list is perhaps the best place to start, since that relates
directly to the taxonomy of writing systems.

As we already noted, all true writing systems encode phonological information in some fashion, and most taxonomies of writing systems recognize this
point, giving primary attention to the different ways in which phonological
information is encoded. For our purposes it will be sufficient to introduce the
basic terminology used to describe the various systems, which will also serve
as a pretty decent field guide to the range of ways in which writing systems
can function.

To wit:

Syllabaries.   The basic elements represented are syllables. Note that this
    does not mean that every distinct syllable of the language has a separate

symbol: indeed it hardly ever means that. What it means is that each symbol represents a syllable-sized chunk of phonological information, such as V, CV, CVV, CVC and so forth. Here by 'V' we mean a vowel such as the /ɪ/ in *hit*, 'C' is a consonant (e.g. the /h/ in *hit*) and 'VV' denotes a long vowel (in English an example would be the /a/ in the German-origin family name *Hahn*). Collectively, consonants and vowels are termed *segments*, a term we will see throughout this book.

Examples: Linear B, Chinese (see more below), Cherokee, Vai syllabary, Yi syllabary, inter alia.

A subtype of syllabary that is often recognized is so-called *moraic* systems, of which Japanese Kana is a prime example. The term *moraic* comes from the phonological term *mora*, which denotes a basic unit of phonological timing. In languages that are analyzed phonologically as having mora timing, a syllable with a short vowel (e.g. CV) counts as one mora, one with a long vowel (CVV) or a vowel and a coda consonant (CVC) comprises two morae. Kana symbols basically represent V or CV units, thus single morae. But Japanese has syllables that are more complex than this, including ones with long vowels (CVV), and ones with coda nasals (CVN), as well as coda consonants (CVC) when the following syllable begins with the same consonant (e.g. *teppan* 'grill'). Representing these requires additional signs, as do post-syllable-onset glides (e.g. *kyō*).

Alphabets. In the literature on writing systems, the term *alphabet* is often reserved for writing systems where each symbol basically represents a single phonological segment, and all segments—both consonants and vowels—are represented. Examples: Greek, Etruscan, Latin, Cyrillic, Armenian, Hangul.

Abjads. Also called *consonantal alphabets*. In these systems the basic elements represent consonants only, though in all modern abjads, some of the consonants have taken on secondary functions for representing some vowels. However, in a pure abjad, such as the original Sinaitic script, no vowels are represented. Abjads also often fail to represent some other important phonological properties, such as for example consonant doubling (gemination). In later abjads, such additional phonological information can usually be represented via diacritics, but unlike diacritic accents in writing systems like French or Spanish, these diacritics are *not* normally used in the orthography: for example, Modern Hebrew text typically contains very few such diacritics. Examples: Phoenician, Hebrew, Syriac, Arabic.

Abugidas. Also called *alphasyllabaries*. In these systems the basic symbols, like those of an alphabet, represent segments, but with two important differences. First, a distinction is made between consonants and vowels in that consonants are typically written inline, but vowels and other non-consonantal information are written using diacritics. Second, abugidas include one or more *inherent vowels*, which are *not* explicitly written: a consonant with no diacritic is understood as being in combination with this inherent vowel. Various abugidas deviate from the above description, in particular with regard to what is represented with diacritics, and where. For example, some abugidas have separate symbols for vowels that begin a syllable, whereas in CV syllables, the same vowels would be written with diacritics; whereas in other abugidas, all vowels are written with diacritics, and there is a special empty consonant symbol for syllables that begin with vowels. Abugidas also differ on how consonant sequences are represented, with some ligaturing the consonant signs together, whereas in others the initial consonant is written inline, with other consonants being written as subscripts.

Examples: the many Brahmic scripts of India and Sri Lanka (Devanagari, Bangla, Odiya, Kannada, Telugu, Tamil, Sinhala ...), Southeast Asia (Thai, Lao, Burmese ...), Insular Southeast Asia (Old Javanese ...), as well as Ge'ez in Ethiopia and Eritrea.

The crucial property of abugidas—and the reason they are also often called alphasyllabaries—is that the symbols are arranged into syllable-sized combinations. If one has a consonant and a non-inherent vowel, then the consonant symbol will be written inline, with the vowel written as a diacritic above, below, after or before the main consonant. In more complex syllables the same principle holds, with the consonants typically ligatured together or in a subscript relationship as described above, with the vowels and other diacritics written around the main consonant(s). This syllabic arrangement often leads people to confuse abugidas with syllabaries. Since this is an important confusion not to make, we discuss this issue further in Section 2.4.

So much for what phonological information is encoded, but of course as is well known, many writing systems may also encode semantic or lexical information more directly. This is hardly surprising since writing originally evolved from symbol systems that primarily represented meaning. You wanted something to represent an ox, for example on a tablet inventorying various farm animals, and a straightforward representation would be a picture of an ox, or a part thereof. This is exactly what Sumerian ⇨ *gud* 'ox' represented: originally a picture of an ox's head, later flipped 90 degrees counterclockwise. This usage is in principle not tied to any particular language, and the symbol can be said to directly represent the meaning. Such symbols are termed *semasiographic*—'writing meaning'. While writing systems, in order to be fully functional, needed to have mechanisms to represent sound, that did not mean that they gave up their semasiographic elements. But insofar as these signs often shifted from representing general meanings to specific words they became, in the terminology of the study of writing systems, *logographic* ('word writing'). Think of the fact that in English the words *ass* and *donkey* both denote the same animal, but are distinct words of the language. A purely semasiographic system would have a sign that could be read either way. If however that sign came to represent just one of those words (with some other spelling developing for the other), then the sign ceases to be semasiographic and becomes logographic.

To be pedantic, many scholars of writing systems prefer the term *morpho-graphic* insofar as signs that are called *logographic* are typically used to write a particular morpheme rather than a whole word. The morpheme *donkey* also happens to be a word of English, but for example the plural *donkeys* is a different word, and in a mature writing system one might expect the sign for *donkey* to be used to write just the first bit of *donkeys*, with a separate sign being used to represent the plural morpheme *-s*.

All the most ancient writing systems—Sumerian, Egyptian, Chinese, and the Mesoamerican systems—were mixed systems that included both logo-graphic/semasiographic elements and phonographic elements—those that write phonological information. Of the writing systems still in current use, only Chinese, and its derivatives used in Japanese (and to a much lesser extent in Korean), still retains this property in that large numbers of Chinese characters (at least 95% of those ever created) contain a component that indicates something about the meaning and a component that indicates something about the pronunciation. See our discussion in Section 1.2. There is a lot of debate concerning to what extent modern readers of Chinese actually make use of this decomposition of characters, but there is no debate about the fact that as a method of forming new characters the semantic-phonetic compound method was indispensable.

Chinese characters as used in the Chinese writing system, with only a couple of notable exceptions, always represent single syllables, which is why we listed Chinese as an example of a syllabic writing system above. However since they also indicate morphemes, or at least parts thereof, Chinese writing is commonly referred to as *morphosyllabic*.

When Japanese adopted Chinese writing the system underwent a number of adaptations, one of which was that the symbols could be used both to represent words borrowed from Chinese—or constructed on the basis of Chinese morphemes—as well as to represent native Japanese words. See Handel (2019) for a detailed treatment of the adoption of the Chinese script for the writing systems of Japanese, Korean and (older) Vietnamese. Indeed the majority of Kanji in Japanese have two (or more) readings, one of which is typically a native word, the other which is a Chinese reading. To take an example that happens to be bimorphemic, written with two characters 故郷, meaning 'home town' (literally 'old country'), can be read as either Sino-Japanese *kokyō*, or somewhat more poetically as *furusato*, composed of two native Japanese morphemes. Both words are bimorphemic with the same meaning, but are clearly different words in the vocabulary. Japanese writing has in effect come full circle in that it contains elements that are best thought of as semasiographic.

2.3.2 How 'transparently' is linguistic information encoded?

When the Romans first developed their alphabet, based on that of the Etruscans, which in turn was derived from a Western variant of the Greek alphabet, the way in which the alphabet represented the Latin language was for the most part simple. Roughly speaking, each letter corresponded to a phoneme, and while certain phonological distinctions were not represented in the system—most notably vowel length distinctions—it was generally fairly straightforward to reconstruct the intended pronunciation of a word from its spelling, or to determine the spelling of a word form its pronunciation. The system was thus fairly regular.

The Latin alphabet was initially a script used for a single language, but is now, with various additional letters and diacritics, the most widely adopted script in the world. If one looks at modern languages that use derivatives of the Latin alphabet in their writing systems, one sees a great deal of variation in how regular the systems are. Writing systems that are historically relatively recent are often quite regular in the way in which sounds map to letters. In Europe, Finnish, whose orthography dates to the 16th century is notable for its regularity. Latin-script-based writing systems of many languages of Africa, the Americas and Oceania, most of which have been developed within the last couple of centuries (often by missionary linguists), also tend to be highly regular.[3]

On the other hand, many of the older writing systems of Europe deviate more or less from having regular letter-sound correspondences. Examples include the writing systems of French, Danish and most notably English. In the case of English, to some extent this can be explained by the retention of archaic spellings that once were reasonable reflections of the pronunciation of a word but where, because of sound change, this is no longer the case. The retention of a 'silent' ⟨k⟩ before ⟨n⟩ in words like *knee, know, knight* and *knave* is an instance of this. Additional factors included:

- The creation during the Enlightenment period of 'etymological' spellings such as *debt*, where the silent ⟨b⟩ was added to reflect the fact that the word originally derived from Latin *debitum*; as well as pseudo-etymological spellings such as the silent ⟨s⟩ in *island* (Middle English *iland*), on analogy with the phonologically similar but etymologically unrelated word *isle* (from Latin *insula* via Middle French).
- And, especially within the last couple of centuries, the adoption of largely unchanged spellings from foreign writing systems (or transliterations thereof), creating pronunciation doublets such as ⟨mole⟩ (insectivorous mammal versus Mexican sauce) or ⟨sake⟩ (benefit versus Japanese brewed beverage).

---

[3] But note that it is not always the case that recently developed systems are regular. The spelling system for Manx Gaelic was developed in the 18th century by a missionary linguist, but is notably irregular in its spelling system.

Psycholinguists who study reading often characterize the difference between, say, Finnish orthography with its highly regular letter-sound correspondences, versus the highly complex system of English in terms of *shallow* versus *deep* orthographies. In a shallow orthography one can, as largely was true in the case of Latin orthography, reconstruct the pronunciation from the spelling, or predict the spelling from the pronunciation. Put another way, in a shallow orthography, one does not need to know the language in order to know how a word should be pronounced: one just needs to know the letter-sound correspondence rules and of course how to pronounce the sounds of the language. A non-speaker of Finnish, who has learned Finnish spelling rules and who has also learned how to pronounce Finnish phonemes, can in fact do a passable job of reading Finnish words aloud without knowing what they mean.

In contrast, this is clearly not generally possible for English. While the basic spelling-sound correspondences might allow one to correctly pronounce *knight* without actually knowing that word, they cannot predict the correct reading of *island*, or *debt*: the fact that the ⟨s⟩ or ⟨b⟩ are not pronounced in these words follows from no general principles of English spelling. One simply has to know the word and the spelling that is associated with it.

In summary, a shallow orthography is one where knowledge of lexical information is in principle not required in order to know how to read a word aloud; whereas in a deep orthography, one does need to know what particular word is intended in order to know how to read it.

A common (mis)conception about writing is the assumption that the 'ideal' writing system is the simplest—simplest in the sense that it has a regular correspondence between letters and sounds, and thus is shallow in the sense described above. Whatever the merits of that view, it is largely beside the point when it comes to logographic or semasiographic elements in writing systems, which by definition do not directly represent sound in the first place. In a logographic system, each symbol ideally represents a morpheme. An ideal pure logograph represents a particular morpheme without any reference to sound. The Chinese characters 人 *rén* 'person' or 馬 *mǎ* 'horse' are pretty good examples. Insofar as they represent morphemes rather than sounds, psycholinguists typically think of logographic systems as instances of deep orthographies, but this is a somewhat different sense of 'deep' than the one that comes into play when one considers a spelling like English ⟨island⟩, since the mapping between symbol and sound is arguably less direct. To read a written form in a logographic system, one must know what morpheme is intended by a given symbol, and how that morpheme is pronounced. Conversely, in order to spell a given morpheme, one must know the symbol that is associated with it.

Pure logographic systems where every morpheme has its own symbol, do not exist. All logographic systems evolved methods to represent sound to greater or lesser extents, so that logography is really a matter of degree rather than representing a categorical taxonomic difference in writing systems (Sproat & Gutkin, 2021). In Chinese, most characters, while representing individual morphemes, also indicate something about the pronunciation of the word. The example of 鯉 *lǐ* 'carp' (= ⟨FISH⟩+⟨lǐ⟩), was given in Section 1.2. Unfortunately, though, while the introduction of a mechanism for representing sound was critical in the development of the Chinese script, it fails to make the system more shallow merely by introducing phonological information. The case of 鯉, where the phonetic piece 里 *lǐ* happens to have the identical pronunciation (in Mandarin) to that of the target morpheme 'carp', is relatively unusual. In most cases the 'phonetic' component of a character only gives at best a hint at the pronunciation. Often this is due to phonological changes that have taken place between older forms of the language that were in use when the system was developed and modern Chinese languages, but in fact the representation of sound in the system was never perfect (Baxter, 1992). More generally, 里 is used as a phonetic component in characters representing morphemes that have pronunciations *similar to lǐ*, and thus could be more accurately described as representing a neighborhood of similar sounding syllables. Furthermore, there is typically more than one phonetic component that could be used to represent a given neighborhood. The upshot is that one just has to know that 鯉 is the character used for the morpheme meaning 'carp'. While the writing of that morpheme 'makes sense' once you know it, there was no way to predict it beforehand from general principles.

Adding to the complexity, most logographic systems do not have a simple one-symbol-one-morpheme structure. In Sumerian, for example, most logographic symbols could actually be used to represent several different morphemes with radically different readings, in addition to their use as purely phonetic symbols. Thus ◁⊣, originally a picture of a head with a depiction of the mouth, represented, among other things, *gu* 'say', *enim* 'word, command', *kag* 'mouth', *zu* 'tooth', as well as serving as a phonetic sign /ka/. Conversely, the same morpheme could often have multiple possible spellings.

Among modern writing systems, the most extreme instance of this is Japanese, where in general the same word or morpheme may be written in multiple ways, and the same character may represent different morphemes. Thus one has to figure out from context whether 東 'east' is to be read as the native word *higashi* or as the borrowed Sino-Japanese reading *tō*; or whether 辛い (Kanji 辛 plus Hiragana い) is to be read as *tsurai* 'painful' or *karai* 'spicy'. Conversely, for the word *tomaru* 'stop' one needs to decide whether it means to stop (e.g. at a stop sign) 止まる; to stop at a station (e.g. a train) 停まる; or to stop at a place (i.e. 'stay') 泊まる. Etymologically, these are all the same word, but in this case the Japanese writing system separates out the senses with different written forms.

But it is worth noting that one finds such issues in writing systems that are not prima facie logographic like Sumerian or Japanese. The most well-known instance is the alphabetic writing system of English, which is famous for having many homophones—words that sound the same but are spelled differently, as well as homographs—words that have the same spelling but represent different morphemes, often with different pronunciations. As with the cases from Sumerian or Japanese cited above, one has to know whether one is talking about music or fish to know how to read the word spelled ⟨bass⟩. Conversely, if one hears /beɪs/, one needs to know whether or not one is talking about a musical range or instrument versus, say, a location for military personnel, in order to know whether to spell it ⟨bass⟩ or ⟨base⟩. For these reasons, the English writing system has often been characterized having logographic properties (Sampson, 1985; Sproat, 2000; Sproat & Gutkin, 2021), despite using an alphabetic script.

Why do we care about these sorts of issues in a book whose primary purpose is to elucidate the complexities of putting what one wants to say into graphical form? There are two reasons. First, to the extent that deeper or more logographic writing systems require more memory and understanding of the language, arguably they also require more mental resources. At least if one wants to write a message that conforms to the conventions of the writing system of one's language, one has to monitor one's output in order to make sure that one does not, say, write *route* when one should have written *root*.

Beyond this, a complex relationship between written and spoken language becomes a technological issue for scripts where the set of symbols does not conveniently fit on a keyboard. Input systems for Japanese, which are invariably based on typing in Kana, and converting where needed into Kanji, ideally must be smart enough to figure out which of the senses of *tomaru* the writer intends. Inevitably an automated system will fail to be 100% correct, and will have to include an option for the writer to select a different output from the one proposed by the system.[4] This of course works—tens of millions of Japanese speakers use such systems on a daily basis—but it does require constant monitoring of the output while one is typing in order to minimize the number of spelling errors. This in turn increases the cognitive load of typing in Japanese, or in Chinese, as Hannas (1997) discusses at length.

2.3.3 How the basic symbols of the script are arranged

Unlike speech, which unfolds sequentially over time and is thus one-dimensional, writing typically occupies a two-dimensional surface. Whether it be a clay tablet, a papyrus scroll, the face of a stele or a piece of paper, the writing surface has a height and a width. Indeed, if one considers cases where writ-

---

[4] In Japanese it is in fact technically *impossible* for an automated system to be 100% correct, since even if it could correctly predict the spelling of ordinary words all the time, there is still the issue of personal names, particularly given names. If one has a friend named *Miko*, one simply has to know how she writes her name: there are several possible ways in which that name might be written, but only one is correct for a given person, and there is no way to predict this on any general principles.

ten symbols are impressed, incised or embossed, there is a third dimension of depth.

And yet while scholars do indeed talk about writing being "two dimensional" or even "three dimensional" (Ferrara et al., 2024), and while such a characterization often makes for a good soundbite when reporting research in the popular science press, the facts are rather more prosaic.

In incised, impressed or embossed texts, writing makes no *significant* use of the third dimension if by 'significant' one means that the third dimension is used systematically to distinguish between signs.

When it comes to two-dimensionality the details are rather more subtle. Significant use of the second dimension in written texts certainly occurs in poetry, tabular arrangements of text, and in other cases where the second dimension has a meaningful function; see Section 3.3, and for more extensive discussion Sproat (2023), in particular Chapter 4. But for ordinary prose text, text is basically linear, with deviations occurring only very locally, typically within glyph combinations corresponding roughly to syllables (Sproat, 2000).

What do these local deviations look like? A figure comparing examples will come in handy, so consider Figure 2.1, which shows the written form of the words for *chicken* in a variety of languages. In this diagram, English presents the simplest case in that the letters are arranged in a strictly left-to-right and thus linear fashion. But consider the other examples, where things are a little more complex. In (b), the Hindi word *murgī*, written in the Devanagari script, shows various deviations from the strictly linear. The alignment between the Devanagari glyphs and their corresponding sounds are shown by color coding, and one can see that the /u/ of the first syllable is written underneath the letter for /m/, which itself is written inline. The second syllable's /g/ is written inline, as is the vowel /ī/. But the /r/, which phonologically is really part of the first syllable, is written as a diacritic on the top righthand edge of the second syllable: as we will see in Section 2.4, while abugidas like Devanagari are often mistakenly referred to as syllabaries, one way in which they fail to live up to that term is that the 'syllable' units in scripts like Devanagari are better thought of as *orthographic syllables* since they generally do not correspond to phonological syllables.

The Kannada word for 'chicken' shows a similar pattern, with many of the symbols being written inline, but with the /i/ of the second syllable written as a diacritic that is ligatured to the top of the preceding consonant.

Finally in Hangul we see a more extensive use of two-dimensional layout, with the four letters arranged in a 2×2 block.

But the systems in (b-d) in Figure 2.1 are only two-dimensional at the very local level. Text in these systems, at the macroscopic level proceeds linearly: from left-to-right in the Brahmic scripts, as well as modern Korean, from top-to-bottom in earlier Korean. The systems might therefore be characterized as 1.5 dimensional (Sproat, 2023).

Be that as it may, this local deviation from linearity can have profound implications for technology. For some of these systems, it has proved difficult

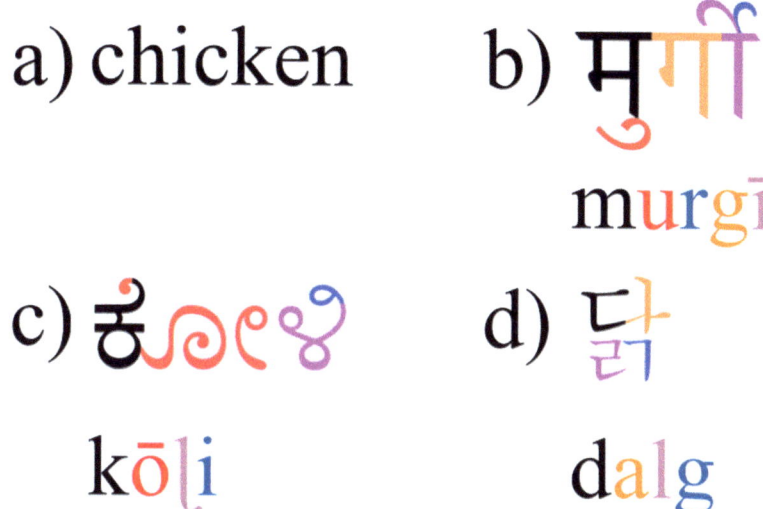

Fig. 2.1 The arrangement of the basic phonographic elements for the word 'chicken' in: (a) the English alphabet; (b) Hindi, in the Devanagari abugida; (c) the Kannada abugida; (d) the Hangul alphabet. The English letters are arranged, of course, in a strictly left-to-right fashion, and the word's spelling is thus constructed strictly linearly. For the other scripts, there are various deviations from a strictly linear arrangement, indicated with colors superimposed on the written forms, with the colors corresponding to the colors of the letters in the romanizations below. For example, notice that the /r/ in Devanagari, though it occurs logically before the /g/ in the Roman transliteration, is written as a hook at the top righthand edge of the sequence of glyphs representing /gī/. Source: own work.

to port a particular input technology; for others it has proven impossible. We explore this issue in detail in Chapter 3.

2.4 What the term 'syllabary' means

As noted above, we do not delve too deeply into issues related to the taxonomy of writing systems, since there have been a great many easily accessible treatments of this issue. However there is one issue we wish to discuss in some detail, namely the term "syllabary", since there is a fair amount of popular misconception about what this means, and understanding how we are using the term is important for understanding issues related to the size of the script, in particular as it relates to traditional printing and typing.

The main point of confusion is between systems where the basic phono-graphic elements of the script represent syllables, and those where the elements are merely *organized into syllables*. In the former case, one generally expects there to be no similarity between glyphs, just because they have phonological elements in common. For example, if a syllabary has symbols for syllables /ka/, /ke/, /ki/, /ko/ and /ku/, there is no expectation that the symbols will have any visual connection with one another.

Such a situation, for example, obtains in the Linear B syllabary, which was used to write Mycenaean Greek. As can be seen in the following, there is no resemblance at all between the glyphs that share the onset consonant /k/:

⊕ ⅏ 𐀒 𐀏 𐀓
ka ke ki ko ku

Now consider the Cherokee script, another true syllabary. Since /k/ and /g/ are not phonemically distinct in Cherokee, we use here the series conventionally transliterated as /g/:

Ꮟ Ꮄ Ꭹ Ꭺ Ꭻ
ga ge gi go gu

Contrast these cases with the situation in Devanagari where the following correspond roughly to the Linear B cases given above, but substituting a long /ā/ for the short vowel /a/:

का के कि के कु
kā ke ki ko ku

In marked contrast to the Linear B case, the Devanagari glyphs all share the basic symbol for /k/, क, which by itself is understood as having an *inherent vowel*, conventionally notated as /a/, so that the whole basic consonant shape represents /ka/. The different vowels are represented by diacritics after, before, above, or below the consonant. This is the pattern found in all scripts derived from the Brahmi script (3rd Century BCE) including most of the scripts of India, as well as many scripts of South-East Asia. It is also the pattern found in the Ge'ez script of Ethiopia, which was probably inspired by Brahmic scripts (Daniels, 2018). The basic consonant shape in such abugidas is augmented or modified by the addition of diacritics representing vowels, and other information.

As another example, consider so-called *Unified Canadian Syllabics* (UCS), used for a variety of aboriginal languages of Canada, and descended from a

script originally invented for Swampy Cree and Ojibwe by the British missionary linguist James Evans in the mid 19th century. Despite the name, this system is again not a syllabary, but an—admittedly unusual—orientation-based
abugida. In the table below, the basic shape of /k/ is the b-q-p-d-shaped
glyph, and the vowels are determined by the symbol's orientation. Note that
there is no /u/ in the basic UCS system:

$$\begin{array}{|llll|}\hline \text{ᑲ} & \text{ᑫ} & \text{ᑭ} & \text{ᑯ} \\ \text{ka} & \text{ke} & \text{ki} & \text{ko} \\ \hline \end{array}$$

Other features are also encoded via further modifications to the basic symbol.
For example long vowels are encoded by a bolder rendition of the symbol:
ᑲ /ka/ versus ᑲ /kā/,[5] among the many features of the script influenced by
shorthand systems, something we will discuss in Chapters 3 and 6.

Or consider another script invented within the last few centuries, Hangul.
In a similar vein to Indian alphasyllabaries, the glyphs are arranged into syllables, consonants and vowels are still separate symbols so that all syllables
beginning with /k/—which is actually a voiceless unaspirated stop initially
despite the Revised Romanization representation with ⟨g⟩—share the same
shape:

$$\begin{array}{|lllll|}\hline \text{가} & \text{게} & \text{기} & \text{고} & \text{구} \\ \text{ga} & \text{ge} & \text{gi} & \text{go} & \text{gu} \\ \hline \end{array}$$

The clear differences between the Linear B or Cherokee cases on the one
hand, and the case of Devanagari and similar abugidas, or Hangul on the
other is somewhat obscured in cases like Japanese *Kana* (both Hiragana and
Katakana).[6] To be sure, like Linear B, the glyphs for syllables that start with
the same consonant generally have nothing in common. Here is the series for
/k/ in Hiragana in the first row, and Katakana in the second:

$$\begin{array}{|lllll|}\hline \text{か} & \text{け} & \text{き} & \text{こ} & \text{く} \\ \text{カ} & \text{ケ} & \text{キ} & \text{コ} & \text{ク} \\ \text{ka} & \text{ke} & \text{ki} & \text{ko} & \text{ku} \\ \hline \end{array}$$

Where the Kana systems primarily diverge from Linear B or Cherokee is in
the treatment of voicing, since this information is marked by adding diacritics to the basic syllable sign. The following are syllables beginning with /g/,

---

[5] In Evans' original system, vowel length was encoded by breaking the symbol.

[6] As noted above, Kana is often more specifically referred to as a moraic system, rather
than a syllabary, though this does not affect the point being discussed here.

where it will be observed that for both Kana systems, the /g/ forms are built from the /k/ forms by the addition of a pair of dots:

が	げ	ぎ	ご	ぐ
ガ	ゲ	ギ	ゴ	グ
ga	ge	gi	go	gu

So it is possible for a syllabary to be either "pure" in that there is absolutely no commonality among syllables that share phonological properties. Or "mixed", like Kana, in that some phonological features are represented in a consistent fashion. But at its core, Kana is still a syllabary in that there is no general way to predict the form in which /Ci/ will be written if one knows how /Ca/ is written.

It is worth pointing out that there is another sense in which abugidas differ from true syllabaries, and that is that the so-called syllables in abugidas are generally not *phonological syllables*, but correspond rather to *orthographic syllables*—termed *akshara* when referring to the Brahmic scripts of India. Consider the Hindi word मस्जिद, representing the Perso-Arabic word *masjid* 'mosque'. If one were to break this sequence down into syllables, it would consist of three pieces: म *ma*, स्जि *sji*, and द *d*. The first is just the symbol for ⟨m⟩ plus the inherent vowel. The third is the final ⟨d⟩, also with an inherent vowel, the latter being elided word-finally in Hindi by a regular rule of *schwa-deletion*. The middle piece consists of the final 's' of the first syllable, ligatured with the initial 'j' of the second syllable, preceded by the short ⟨i⟩, which logically follows the 'j'. The phonological syllabification of the word is *mas-jid*. Yet there are three orthographic syllables, with the second one combining components of both phonological syllables. In a true syllabary, each basic symbol is in fact a single syllable. This does not mean that every phonological syllable has its own symbol: phonologically complex syllables may in fact be written with several symbols in a syllabic writing system. But symbols in syllabaries do not correspond to phonological pieces from separate syllables, as in the case of स्जि *sji* discussed above.

Why does this matter, apart from a general desire to keep terminology precise? One way in which it matters is to learners of the system. If one treats Devanagari as an abugida, then one needs to learn roughly on the order of 120 glyphs, plus rules for their combination and exceptional ligatures. If on the other hand one treats them as syllabaries, then one needs to learn thousands of syllables: for example there are over 540 'symbols' just to cover simple consonant-vowel combinations. This has direct implications for education. In fact, in India it is common to teach the Brahmic scripts as syllabaries. Nag & Snowling (2011) discuss this point in the context of literacy education in Karnataka in the Kannada script, noting that children are introduced to the script as a syllabary, which is one of a couple of pedagogical decisions that appears to slow literacy acquisition.

The question for abugidas and other compositional systems that are sometimes mistaken for syllabaries is whether from a practical point of view it is better to treat them as syllabaries despite their obvious compositionality. Or rather to consider them to be essentially alphabets, albeit ones where the combination rules are more complex than simple linear concatenation. There is no doubt that the combination rules in abugidas need to be learned, and they can themselves be complicated. To take a relatively simple example, consider the ligatured form of 'l' and 'i' in the written form of the Kannada word *kōli* in Figure 2.1c, where the /i/ appears as a small circle attached to the 'l'. The diacritic 'i' generally appears in Kannada script in the form of a small circle, but how it attaches to particular consonants is idiosyncratic and needs to be learned.

It is therefore an empirical question that must be examined on a case-by-case basis whether it is more efficient to ignore the internal compositionality and simply treat the system as a syllabary, with of course a much larger character set. Or to treat only the basic symbols as part of the inventory, but add to that inventory the combination rules and exceptional cases needed to produce well-formed combinations. Which decision one makes is of importance not only to those trying to master the script, but also to the technology required to produce it. As we shall see in the next chapter printing and typing technology has addressed this issue in various ways, depending on the particularities of the given script.

Chapter 3
Script complexity

3.1 Prelude

The act of writing, on any medium, involves implementing a sequence of gestures, which produce or select from a set of basic signs or sign components, which may be assembled into more complex compound signs, which are arranged into text. Any stage of that process may involve some ambiguity as to what was intended. And at any stage of that process, the technology being used to write may make assumptions that limit the flexibility of the system.

Consider two examples, the traditional typewriter, and a modern computer-based input system for Chinese. Writing with a typewriter, involves learning a set of gestures to select symbols by striking a set of keys. Since there was a one-to-one mapping between symbols and keys, or key combinations (e.g. Shift+A), typing allowed one to input text without ambiguity, and a skilled typist could do so without error. But the typewriter was of course limited in various ways. It was limited in the number of keys, so that only scripts with relatively small alphabets were readily supported. And it worked best when the script was strictly linear, so that successive strikes of the keys would produce symbols in sequence across the page. Diacritics such as accents that had to be added to a previously typed letter would slow things down. And some scripts, such as Brahmic scripts or Hangul were never easily supported since they involved many deviations from a strictly linear typing sequence, making it mechanically difficult to design a typewriter that gave pleasing results. Scripts with large numbers of symbols, such as Chinese, were impossible.

With computer input systems, of course, the situation changed radically. The keyboard of a computer is not largely different from that of a typewriter, but now the limitations are not mechanical, but algorithmic. Hangul has a small enough set of basic letters—*jamo*—that it will easily fit on an ordinary keyboard. Algorithms can deterministically convert sequences of typed jamo into syllabic combinations, with the end result that it is just as easy to type Korean on a computer as it is to type English. The removal of mechanical limitations means that Chinese is also straightforward enough, with the one caveat that, while there are a number of different methods available to input Chinese text it is still not possible to type Chinese characters directly. The most popular phonetic (typically *pinyin*) based systems mean that one can type in Chinese using a standard keyboard just as one would type in English, but the output of one's typing must be converted using a language model[1] into the intended sequence of characters. As it happens, the second author's first patent was for a Chinese input system that used a bigram language model to convert from phonetic input to character sequences[2], and while language modeling has vastly improved since that time, the conversion still cannot be done entirely without error. This in turn means that when one types in Chinese—unlike the case of English or Korean—one needs to monitor the text that appears on the screen to make sure that the system did not introduce an error.

There are also input systems, such as *Cangjie*, which is based on a structural decomposition of characters, requiring a great deal more skill to learn than phonetic input methods. The system is largely, though not completely unambiguous: there are cases where the same input sequence will involve collisions between multiple possible characters, and so Cangjie provides a mechanism for choosing characters in such cases. Thus the system still requires some monitoring to make sure that one is getting the correct output.

3.2 Introduction: Forms of complexity in writing

The Japanese writing system is well-known to be difficult to learn. Native speakers of Japanese spend years in school mastering the system and for adult non-native speakers, the system is truly daunting. But what precisely is it about Japanese writing that makes it a complex writing system?

In fact there are several ways in which it is complex. First of all, as most readers will know, the Japanese writing system comprises three scripts: *Kanji*, or Chinese characters; and two syllabaries, *Hiragana* and *Katakana*, collectively just called *Kana*. The latter were actually originally both derived, via different routes, from Chinese characters, but are purely *phonographic*—i.e. they are only used to represent sound. Simplifying somewhat, Kanji, on the other hand, are best described as morphographic, meaning that each symbol (roughly) represents a word or a morpheme. These three systems each have their own function. In Modern Japanese, grammatical morphemes are usu-

---

[1] A language model predicts the next token given previous tokens, and is the means, in this case, to finding characters that are frequently used together. Chapter 6 provides more details on how such models are used for text entry.

[2] US Patent 5270927, September 10, 1990. A bigram language model conditions the probability of each token on just the previous token.

ally (though not always) written in Hiragana. Katakana tends to be used for foreign words and names, names of companies, names of animals and plants, and for emphasis, among other things. Kanji are usually used to write content words, native as well as Chinese or Korean personal names, names of some countries—though all of these may be found written in other ways. Thus for example, *America* is typically written these days in Katakana as ア メ リ カ *amerika*, but it can also be found written as 米国; the latter can also be read as the Kanji suggest as *beikoku*, but it is more typically read as *amerika*. In general one will find exceptions to everything we have just described thus far so that the Japanese writing system is complex not just because it has many rules or principles governing its use, but because those rules or principles are often broken.

The Hiragana and Katakana sets are both relatively compact, but the set of Kanji, on the other hand, is quite large. Thus the list of *jōyō Kanji*, or Kanji in common use, tops 2100, but it must be emphasized that this is just the Kanji that are common and are deemed essential to know. One will often find Kanji outside the list, especially in personal or place names, where rarer Kanji or older variants of Kanji may still be used. Thus the family name *Takimoto* could be written with the *shinjitai* (new form) *jōyō* first Kanji as 滝本, but there are families who still write it with the *kyūjitai* (old form) Kanji as 瀧本.

The larger the number of symbols in a script, the more complex on average the symbols need to be and, contrariwise, the smaller the set, the less complex on average the individual symbols. This is obvious: if one needs to make more distinctions, one needs a larger set of possible forms, which leads to the forms, on average being more complex. On the other hand, if one has a small set of symbols, there is not much benefit in having complex forms, and so the basic symbols will tend to be much simpler. In fact our Roman alphabet is derived ultimately from an early Semitic script (Proto-Sinaitic), where the letters were actually pictures of objects. Thus, ⟨A⟩ comes originally from 𐤀, a line drawing of an ox head, which is at least somewhat more complicated than the letter ⟨A⟩. Kanji, on the other hand, tend to be much more complex in form, making them harder to learn: In contrast Hiragana and Katakana, which were derived from Kanji are much simpler in form since the number of distinctions they need to make is much smaller.

This is the first of two chapters that deals with the question of what makes a script and a writing system more or less complex. While up to this point in the discussion we have been considering the human learner, the focus in this chapter is on the question of what makes a script complex from the point of view of getting text down onto some medium—whether that medium be clay, parchment, paper, or electronic memory. As we will see, the complexity in turn varies with the technology, and there are various tradeoffs. A pen (brush, stylus, ...) has the fewest implications on complexity, at the cost of forcing the writer to remember how to write all the symbols they want to use. Typewriters and Linotype machines impose rather severe restrictions,

making them impractical or at least very difficult for certain scripts. For
typewriters in particular, a reasonably compact keyboard is desirable, and
this limits what can easily be put on the keyboard. Modern computer-based
input systems make inputting text a lot easier, but keyboard restrictions are
still relevant.

In the next section we outline the aspects of *script complexity* that various
input technologies are affected by.

3.3 Three types of complexity

As noted in the previous section, two aspects of the complexity of learning
and using a script are the number of glyphs, and the glyph complexity. These
tend to be inversely related.

A third aspect is the script's *linearity*. In all mature writing systems, text is
basically linear. The text may run from left-to-right, or right-to-left, or top-to-
bottom, but it generally does so by arranging the symbols in a line: writing is
essentially one dimensional. True, the physical page is two-dimensional. And
in poetry, for example, line breaks often play a significant role. Significant
use of the second dimension is particularly used in so-called *visual poetry*—
Figure 3.1. Similarly, in tables, two-dimensional arrangements are important,
and of course when text is integrated into artwork—monumental inscriptions,
or comics—the second dimension is often critical. But in ordinary prose text,
line breaks and page breaks occur because there is a limited area on the page,
and the placement of those line and page breaks generally has no significance.
Thus, writing is effectively one-dimensional. In an English written sentence
the letters are placed o n e   a f t e r   a n o t h e r in a line.

But the characterization we just gave needs to be modified somewhat, since
while writing is always *macroscopically* linear, scripts vary on how the basic
elements are combined at a local level. In English the word *one* is written as
a sequence of letters ⟨one⟩. But in Korean Hangul, as we noted in Section 1.5,
the basic letters are not generally arranged linearly, but rather are grouped
two-dimensionally into syllable-sized combinations. Hangul can thus be said
to be "1.5 dimensional" as we noted previously. While this property does not
notably add to the difficulty of inputting text if one is using a pen, if using
a typewriter, or a Linotype machine, the non-linearity of Hangul becomes a
technological challenge.

There are thus three basic properties of script that can affect how easy or
difficult it is to input text:

- Glyph-set size
- Glyph complexity
- Linearity

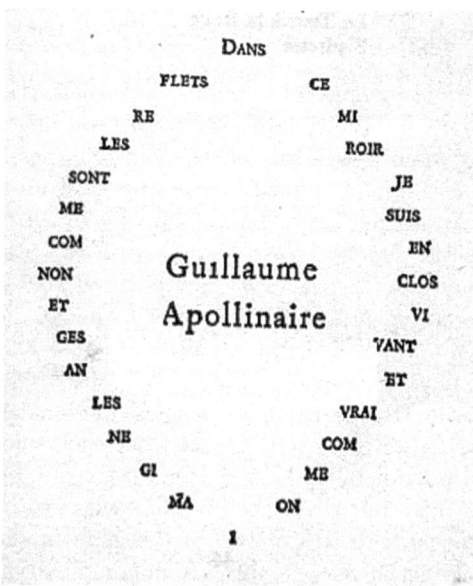

Fig. 3.1 Visual poem in the form of a mirror, Guillaume Apollinaire. The text of the poem forms the frame of the mirror, and the author is "reflected" in the mirror, by means of placing his name in the middle of the frame. Work is in the public domain in its country of origin and other countries and areas where the copyright term is the author's life plus 70 years or fewer.

While in theory the first two properties are independent, in practice they are highly linked. In larger glyph sets, one finds more complex glyphs than in smaller glyph sets, and thus we treat the first two issues together in the next section. Linearity is treated in Section 3.5.

3.4 Glyph-set size and glyph complexity

The five or six-year-old child—mostly boys, but apparently also girls (Charpin, 2023)—who trained to be a scribe during the Old Babylonian period had a long grueling study ahead of him or her. The training took place in the *eduba*, Sumerian for 'the house of tablets'. Since the students by that period would have been Akkadian speakers, part of that education involved learning the Sumerian language, which was still used as the language of administration during the Old Babylonian period (Lucas, 1979; Robson, 2001; Paulus, 2023). But apart from learning a new language, the students also had to master the complex Cuneiform-based writing system for Sumerian, which comprised hundreds of signs. Some of these signs represented whole words, but many also represented syllables, or fragments of syllables. Thus for example in one of the spellings for *eduba* ⟨E₂ DUB ba a⟩ ▰ ▰ ◁ ᛁᛁ, the first two glyphs were *logographs* for 'house' and 'tablet', respectively, and the last were *phonographs* for the syllables *ba-a*. In fact, nearly all glyphs were multivalent, having many readings: five or more readings for any given sign was not at all unusual. Signs ranged in complexity from single impressions of the stylus on clay, all the way to tens of impressions for some of the more complex signs (e.g. the sign ▰ for *ug* 'lion', with well over 20 strokes).

Students in the eduba started with practicing syllabic glyphs. They then proceeded to copying lexical lists, long lists of names of animals, birds, fish, gods, place-names among others (Lucas, 1979; Paulus, 2023). The multiyear program of study obviously involved more than just learning to write: as noted above, the students also had to learn a foreign language—one that by the way was structurally quite different from their own. But there is no doubt that a major portion of the time needed to learn to be a scribe involved simply learning to read and write the glyphs needed to be a master of the system. And this in turn relates to two of the aspects of script complexity listed above, namely glyph-set size, and glyph complexity.

Shift to the modern era, and students in China and Japan similarly spend years learning their writing system. Japanese children start learning kanji in their first year of elementary school. They are expected to learn to read and write the 1,006 characters in the *Education Kanji* list by the end of sixth grade (Halpern, 2006), and the complete set of 2,136 (*jōyō*) ('common use') Kanji by the end of ninth grade[3]. This would also involve learning the various readings of these Kanji.

Obviously Japanese children study other things during those years besides reading and writing, but contrast this with the situation in Finland. Finnish has an alphabetic writing system based on the Latin script, with a few extra diacritics, and it has a very regular correspondence between letters and sound. As a result about one third of Finnish children are *already reading* by the time they start school at age 7.[4]

While glyph-set size is in theory orthogonal to glyph complexity, in fact the two are in practice highly linked. Obviously the more distinctions one needs to make in the system the more one needs to have ways to make the signs visually distinctive. In theory this could be done without significantly increasing the complexity of the glyphs. To see this consider again James Evans' script for Swampy Cree which, in an expanded form used to write a variety of Canadian Aboriginal languages, goes by the name of Canadian Aboriginal Syllabics. Evans' system, originally developed for Cree and subsequently Ojibwe (both Algonquian languages), and inspired in part by Brahmic scripts like Devanagari, but also Pitman shorthand, had an innovative way of representing vowel distinctions in CV syllables. Each consonant had

---

[3] https://en.wikipedia.org/wiki/Kanji.

[4] https://thereadingadvicehub.com/early-literacy/finland-literacy-rate/

	grapheme	name	gloss	strokes
simple grapheme	⟶	$y\bar{\imath}$	'one'	1
complex grapheme	䨻	$b\grave{e}ng$	'thunderclap'	52

Table 3.1 Examples of the simplest and one of the most complex Chinese characters, with stroke counts.

a particular shape, and the vowels were represented by changing the orientation of the consonant glyph. For example, the syllable /ti/ is written as ∩. The same glyph shape pointing downwards ∪ represents the syllable /te/. Pointing left ⊂ it represents /ta/, and right ⊃ /to/. This orientation-based system is used consistently for other consonants. Additional distinctions such as vowel length are marked with other devices: in Evans' original system by breaking the glyphs, in modern versions by bolding or diacritics. The system is thus not really a syllabary but rather, like its Brahmic script inspiration, an *abugida*, but one where the basic vowel distinctions are marked by orientation rather than diacritics.

Suppose one were to adapt Evans' approach of using orientation to mark distinctions more generally? Evans' orientation distinctions were restricted to 90° changes, but in theory one could introduce more fine-grained orientation differences. If the basic set of glyphs contained 50 symbols, one could expand that to 5,000 distinctions simply by allowing for rotations of 3.6°, without any increase in the complexity of the glyph set. Of course, such a system would be of little practical use: it would require supremely fine-grained control on the part of the writer, and significant visual acuity on the part of the reader.

Thus, invariably, the more distinctions a script needs to make, the more complex on average the signs in the script need to be. Thus between the two graphemes shown in Table 3.1—a simple one with a single stroke and a very rare complex one with 52 strokes—there is a large range of complexity in the thousands of Chinese characters found.

Traditional writing implements and writing surfaces were largely indifferent when it came to glyph set complexity—as well as linearity, as discussed in the next section. Assuming one had enough skill and a fine enough implement, one could create glyphs with a wide variety of strokes and arrange those strokes and other basic script forms in various ways on the surface. One could, for example, design one's script so that consonants were written largely inline, but vowels were written as diacritics above, below, to the right, to the left or even surrounding the consonants: many Brahmi-derived abugidas have exactly this property.

To be sure, properties of the medium had an effect on the outward appearance of the script. When Chinese writing shifted from writing with pens on bamboo strips, to writing with ink brushes on paper, the form of the script naturally changed. The Brahmic scripts of India were traditionally written on palm leaves, but with the difference that in the north one wrote with a pen on the leaves, whereas in the south a metal stylus was used to incise the glyphs on the leaves, to which an ink was later applied to make the incised forms more legible. Palm leaves cut easily if one incises in straight lines, and so to avoid this, scribes using metal styluses tended to use more curvy forms. The result is that the southern scripts of India are notably more curved in form than the scripts of the north (Singh, 1986). As a final example, the wedge-shaped basic strokes of the Cuneiform family of scripts came about because it was found that it was easier to mark clay with a stylus by pressing the end of the stylus into the clay, than to incise figures, as had been done in the early pre-Cuneiform scripts. But within these sorts of limits, one still has a great deal of freedom with a pen, brush or stylus. Thus traditional writing tools are largely indifferent to how many glyphs there are in the system.

On the other hand glyph-set size matters a great deal if one is trying instead to fit the script onto a keyboard.

The Latin alphabet of twenty-six letters, or fifty-two if you include capitals, plus ten digits and a few handfuls of punctuation symbols fits easily onto a standard typewriter keyboard. Of course this is in itself no surprise: the modern keyboard evolved out of the first typewriters developed by Christopher Latham Sholes for Remington, and these were developed in the United States, with English as the target language. Sholes' invention had many predecessors dating back to the 18th century, but all of these were for languages that used alphabetic script-based writing systems with small numbers of characters. Selected conventional typewriter history will be presented in Chapter 5, and see Beeching (1990) for a comprehensive history of the development of typewriters.

For most users, the typical typewriter keyboard hits a "sweet spot" when it comes to the number of keys that one must master. But as we will see in this chapter as well as later in the book, there are various limitations. For traditional typing (as well as printers such as Linotype machines that also use keyboard input), there were constraints on what scripts could be handled. And then there are users who, due to motor or cognitive disabilities, may not be able to control a regular typewriter keyboard—a topic we take up in Chapter 5.

The invention of the typewriter benefited greatly from the relatively small number of glyphs in the Latin alphabet, and the fact that the letters are linearly arranged—English writing is 1D not 1.5D. Indeed, whether one can fit the basic set of glyphs of a script on a keyboard is therefore only part of the issue: equally important is whether the technology can combine the typed glyphs correctly according to the way the given script works. That relates to linearity, the topic of the next section.

3.5 Linearity

As noted above, in many scripts the basic symbols can be combined in ways more complex than simply stringing them together in a line. In Hangul, for example, orthographic syllables are composed of individual letters—*jamo*. The orthographic syllables are written in a line, but within the syllable each jamo is written either to the right or below the jamo it logically succeeds. Thus while the *macroscopic* direction in Modern Korean is left-to-right, within the orthographic syllables, the direction of the symbols deviates from this macroscopic direction. In the Brahmic scripts of India, even more complex deviations from the macroscopic left-to-right direction can be found, with some symbols being written to the right, left, above, below or even surrounding the symbols they logically follow.

Thus there are many scripts where the basic elements that can be combined locally in directions that deviate from the macroscopic direction. Since this description is rather a mouthful to repeat, we will designate such scripts henceforth as **non-linear scripts**.

The important point with non-linear scripts is that while the basic inventory may be small, the ways in which the symbols combine can be problematic for some technologies. There are twenty-four *jamo* in Hangul, which means that the set can easily fit on a typewriter keyboard, or on the keyboard of a hot-metal printer such as a Linotype machine. But, as we will see below, the non-linear ways in which the letters are combined in Hangul have proved challenging for various text input technologies. As we document below, Linotype was basically impossible if one wanted to print text horizontally. And while typewriters were developed for Korean, the technology was not trivial to develop, and the resulting output was often rather ugly. So much so, in fact, that there was a movement to simplify the script to make it easier to adapt to the technology. A similar set of issues arose for the Brahmic scripts of India, and in that case there actually were script simplifications motivated by the move to modern technology. The story behind printing and typing for non-linear scripts is presented below in Sections 3.6 and 3.7.

3.6 Printing and script complexity

The use of embossed signs and images to make impressions on surfaces is a technology dating back thousands of years. Square and cylinder seals were used in the great bronze age civilizations of Mesopotamia and the Indus Valley to impress clay surfaces with images, symbols and text; see Figure 3.2 for an example of a Sumerian cylinder seal and its (modern) impression.

The earliest example of 'movable' type (as pointed out by Fischer, 1997) may well be the Phaistos Disc (Pernier, 1909; Evans, 1909), a 2nd millennium BCE artifact from Phaistos, Crete. The Disc is impressed on both sides with

Fig. 3.2 A Sumerian cylinder seal of Queen Puabi, 2,600 BCE Ur, along with a modern impression. The seal contains the Cuneiform text 𒁀𒉿 𒊩 *pu-abi nin*, roughly translating as 'queen/lady Puabi'. Source: By Nic McPhee from Morris, Minnesota, USA—British Museum with Cory and Mary, 6 Sep 2007 - 185, CC BY-SA 2.0, `https://commons.wikimedia.org/w/index.php?curid=10605639a`.

a text consisting of pictographic glyphs, where each instance of the identical glyph was clearly printed with the same stamp.[5]

3.6.1 Woodblock printing and movable type

However what we normally think of as printing involves the use of an embossed surface that has been inked, and where the inked surface is imprinted upon a lightweight material such as paper. That technology, as is well known, was invented in China during the Tang Dynasty around the 7th century AD (Tsien, 1985). An artisan would carve a text on a wood block, and then that wood block could be used to print multiple texts. One of the most impressive examples is the *Tripiṭaka Koreana*—Korean 팔만 대장경 (八萬大藏 經) *palman daejanggyeong*—carved in the 13th century, consisting 81,258 woodblocks, and representing the oldest surviving example of Buddhist canon in Chinese script.

Woodblock printing is more or less as "script-indifferent" as the brush or pen. The only factor that is relevant in transferring the text to the woodblock is the skill of the artisan.

The invention of movable type[6] using porcelain, in the first half of the 11th century during the Northern Song Dynasty in China (Daniels, 1996) introduced a technology where the number of glyphs in a script, and their mode of combination, became important. It is therefore somewhat ironic that

---

[5] Since the text of the Disk has never been convincingly deciphered, despite many attempts, it is as yet unclear whether the text is in fact writing.

[6] Or its rediscovery if we count the Phaistos Disk mentioned above

it was invented in China, where the script did not allow for a small number of types.

Movable type introduced two technological innovations. The first, obviously an advantage over woodblock printing, was that one merely had to produce enough type to cover the entire set of characters in the script, and make sure that the more common characters had enough type instances to cover their expected frequency on a typical page of text.

The second technological innovation was a direct consequence of the need to anticipate how many instances of a given character would be needed to set a single page: the earliest practical application of statistical natural language processing was computing the frequencies of characters so that a printer would be have enough instances of a common character like 也 *yě* (a frequent sentence final word in Classical Chinese), but could afford to have few instances of a relatively unlikely character like 龜 *guī* 'turtle'.

The earliest extant book printed in Chinese in movable type is again a Buddhist text from Korea, the *Jikji* (직지, 直 指), printed in 1377.

The set of type needed for Chinese was quite large, requiring printers to develop good memory skills in order to be able to remember where in the cases a desired character could be found. When Gutenberg invented metal movable type around 1450, he could immediately take advantage of the much smaller character set of the Latin alphabet.

When Hangul was invented in Korea in the 15th century, the simple alphabetic script could in principle have led to a simplification of printing using movable type. Yet the decision to group the individual letters into syllable blocks mitigated against this. In theory the printer might have developed separate type for each *jamo* letter and then assembled them into syllable blocks, along the lines of what was later done with Korean typewriters (Section 3.7.1), or what was done in Brahmic-script movable type (Section 3.6.4). But this would have been an exceedingly fiddly process, and in fact what was done was to produce type containing precomposed syllable blocks. Since this involved at least 2,000 blocks for commonly used syllables, and perhaps a few thousand more for rarer ones, the complexity of Hangul movable type printing was not notably different from that of Chinese. In fact the situation would be rather worse, since up until the late pre-modern period, mixed texts consisting of Hangul, and Chinese characters for words of Chinese origin were the norm (see Figure 3.3), meaning that the compositor would need to deal with two large sets of type.

As a practical matter, this was not really an issue, since despite the fact that movable type had been known in Korea since the 14th century, woodblock printing was still largely the technology of choice: The *Hunmin Jeong'eum*, the document produced under King Sejong's direction that promulgated the Hangul script, was woodblock printed, for example. Movable type printing for Hangul was not introduced until the late 19th century, and then largely for missionary activities, with type fonts produced in Japan (Park, 2011). Furthermore, printed books in Hangul were not particularly

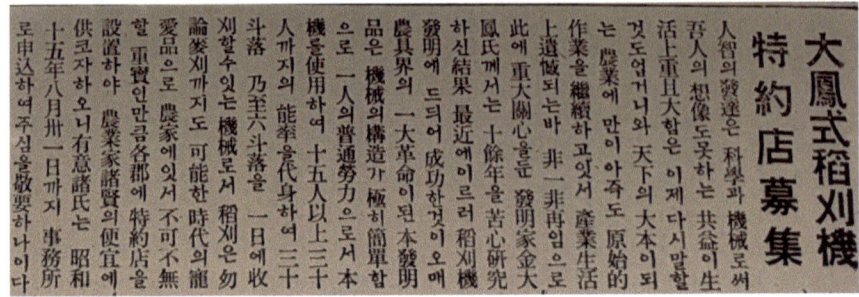

Fig. 3.3 . An example of mixed Hangul and Chinese characters (*Hanja*) in an article in the *Chosun Ilbo*, August 12, 1940. Source: own work.

needed: official court documents, and documents aimed at the scholarly class continued to be written in Chinese, and the printing of Hangul books was limited to a few types of text (see Section 7.4). The technological problem that Hangul presented for movable type technology was effectively deferred for a few centuries.

3.6.2 Monotype

Between Gutenberg and the late 19th century there were many innovations in the history of printing. Notable among these were the invention of lithography by Senfelder in 1796, Konig and Bauer's invention of the "perfecting press", which allowed for printing on both sides of a piece of paper, and David Bruce's invention of the pivotal typecaster, which drastically sped up the production of type production (Timmers, 1987). But it was the invention of hot-metal printing that most significantly changed the course of printing prior to the 20th century.

Though invented in 1885, a year after Linotype™ (Section 3.6.3), Monotype™ is technologically prior in that it is closer in spirit to movable type. Unlike Linotype, which used *hot metal* technology to cast slugs for a whole line of type, Monotype cast single character types, which were then inserted into the typecarrier.

The Monotype machine was actually two machines. The first was a keyboard on which the operator would enter the text to be printed. As a key on the keyboard was depressed, it would cause a line to be punched on a paper tape (called a 'ribbon'). Once the full text was entered, the tape reel was removed from the tape holder, and at that point the keyboard operator could write additional instructions (e.g. which particular typeface was to be used) on the tape, which was then passed to the operator of the *caster*. The caster was a hot-metal casting device that used a *matrix* of character molds

Fig. **3.4** A Monotype matrix case. Source: `https://en.wikipedia.org/wiki/` `Monotype_system#/media/File:Matrixcase-bembo-16pts.jpg`, Author: Enkidu1947. Image is in the public domain.

whose selection was controlled by the perforation patterns on the tape reel produced by the keyboard operator. The original matrices were 15x15, meaning they could hold up to 225 characters. See Figure 3.4 for an example of a Monotype matrix case.

Whereas the Linotype operator controlled everything from inputing the text to casting the slugs, with Monotype there was a division of labor between the keyboard operator and the caster operator. The keyboard operator was responsible for entering the text, and writing instructions on the resulting tape roll to specify things like font size and style. The caster operator produced the actual type, paying attention to the written instructions from the keyboard operator, and selecting the appropriate matrix for, say, 10 point italic, or whatever was specified.

Monotype had a few advantages over Linotype that were important for the technology's adaptation to other scripts. First of all, it allowed for kerning, meaning that characters could overlap, something that was impossible with Linotype. Second, while the number of characters was limited by the matrix size, which in turn was limited by what the Monotype caster could accommodate, there was nothing in principle preventing one devising a machine based on the same general principle, which could accommodate a much larger matrix. Indeed, Japanese manufacturers built 'Monotype' machines that allowed for matrices with over 2,000 characters—see Figure 3.5—and there were several models that supported the post-war newspaper industry in Japan, as well as Korea. Support for such a large character set would have been impractical for Linotype.

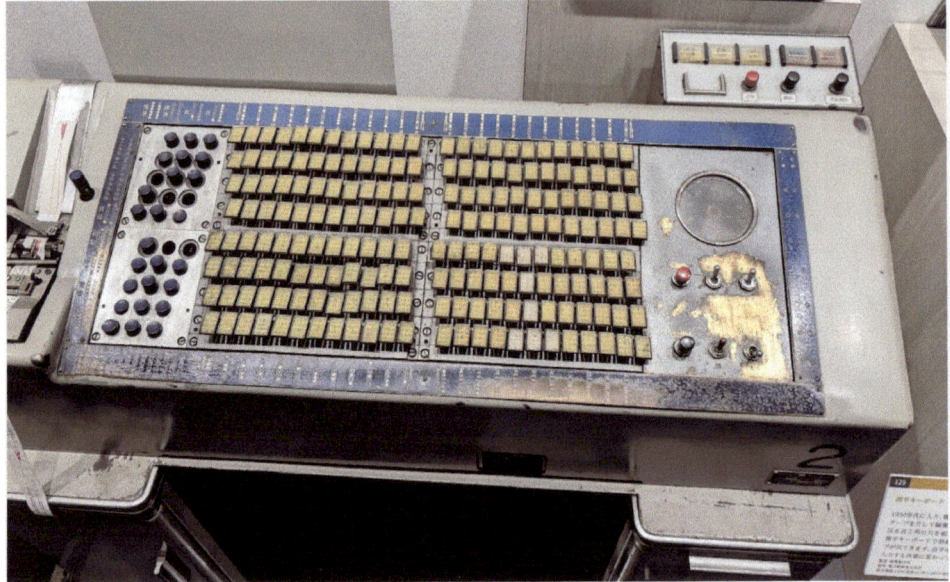

Fig. 3.5 Nippon Type MTH 2304-key 'Monotype' keyboard and tape punch used by *Asahi Shimbun* from 1959 onwards. Each of the 192 keys on the keyboard handled 12 characters, the selection of which involved the additional buttons on the left of the keyboard. The system allowed for an input speed of up to 103 characters per minute (CONPT, ND). In the collection of the Newspark Museum, Yokohama. Source: own work.

3.6.3 Linotype

In the Season 4 (1963) episode of *The Twilight Zone*, "Printer's Devil", Douglas Winter (Robert Sterling), the editor of a failing small town newspaper is approached by a mysterious "Mr. Smith" (Burgess Meredith), who offers his talents as both a reporter and Linotype operator to help Winter turn around the paper's fortunes. Smith, who it is soon revealed is Satan, introduces some "modifications" into the paper's Linotype machine that cause any story that is entered into the machine to become fact. While this modern-day version of the Faust story was of course fantastical, the real life Linotype machine did in fact change the world.

Indeed, the invention of the Linotype machine in 1884 by Ottmar Mergenthaler (1854–1899) was a revolution in printing. Prior to Linotype, movable type printing involved cases containing character blocks ('sorts'), from which the printer would select characters in sequence. While printing with a set of individual type pieces was a fairly script-indifferent technology—with some caveats, a point we will return to below—traditional hand-composited movable type had the disadvantage that it was slow.

Now, the process of typesetting could be mechanized. The cases could be replaced with *magazines* containing type, and the selection of type controlled by a keyboard, and this could certainly speed up the process. According to Howe (1947) (pp. 62–63), typesetting machines allowed for the production of at least 6,000 ens per hour (an *en* being half the body height of the typeface), or about three times the output of a fast hand compositor. However, Howe notes that the machines were never really economical unless they could be operated by cheap labor.[7]

Linotype changed all that.

Linotype was a *hot metal* printing technology that assembled character molds, called *matrices*—*mats* for short—into the *assembler block*, which functions as a mold for a whole line of type—hence the brand name. These molds were then injected with a liquid alloy of lead, antimony and tin heated to a temperature kept between 535 and 550 degrees Fahrenheit (Mergenthaler Linotype Corporation, 1940, page 177). Once the *slug* cooled and solidified, it was assembled into a *galley tray*, which accumulated the lines of text—say for a column of a story in a newspaper. The machine was equipped with 90 keys, which controlled the emission of mats from the magazine. The keyboard contained lower-case letters (on the left-hand side), upper-case letters on the (right), numbers, punctuation and special characters. Spaces are inserted using a *spaceband lever* to the left of the keyboard, which causes *spacebands* to be inserted into the assembler block. Additional characters could be selected manually from *side sorts*. Different typefaces were arranged in different magazines, so that changing typefaces simply meant changing to a different magazine, and later models enabled switching typefaces within a line by causing the machine to select from the desired magazine for each portion of the line. Still later machines allowed for the control of the keyboard with a paper tape, meaning that the text could be pre-typeset, and then simply fed through the machine to produce type. Figure 3.6 shows the 90-key keyboard and spaceband, and Figure 3.7 shows an example of the assembler block with matrices. The Linotype was an incredibly complicated device with thousands of parts—famously called by Thomas Edison the "eighth wonder of the world" (Wilson, 2012).

The technology transformed the printing industry, so that newspapers could be produced much more rapidly and in much longer editions. It greatly lowered the costs of producing books and magazines, and thus ushered in the age of mass media. Larger printing houses often had dozens of linotype machines, employing a corresponding number of operators: the *New York Times* had 140, the largest collection in the world (Halff, 2023).

---

[7] In the infamous case of the Paige Compositor, in which Mark Twain invested, and lost, a great deal of money (Goble, 1985), the machine had the double misfortune of having numerous mechanical problems, and appearing on the market at roughty the same time as Merganthaler started marketing the Linotype.

Fig. 3.6 The Linotype keyboard, spaceband to the left Source: `https://commons.` `wikimedia.org/wiki/File:Linotype_machine_keyboard_at_the_Washington_Post_-` `_DSC05530.JPG` License: CC0 1.0 Universal Public Domain Dedication.

3.6.4 Monotype, Linotype and script complexity

3.6.4.1 Brahmic scripts and Linotype

Linotype was of course first developed for English but systems for many other languages were developed. The Linotype corporation listed fonts for, besides Latin script, Greek, German Fraktur, Cyrillic, Hebrew, Arabic, Armenian, Syriac, Bangla, Gujarati, Tamil and Devanagari (Mergenthaler Linotype Corporation, ND, 1933; Govil, 1937). Of the scripts listed above, Latin, Greek, Fraktur, Cyrillic and Hebrew all have relatively small character sets, and for the most part can be printed without ligatures. These are natural fits to the Linotype technology. One might wonder whether Hebrew, which runs from right-to-left, would necessitate a redesign of the assembly mechanism, but in fact it did not: one could employ the simple expedient of putting the characters on the matrices upside down (Halff, 2023), so that the characters get assembled left to right into the assembler block the same way as when typesetting English, but with the line upside down. Arabic and Syriac are cursive scripts, meaning that many characters *must* be joined together, but this can be achieved by having variant joining and non-joining forms, which

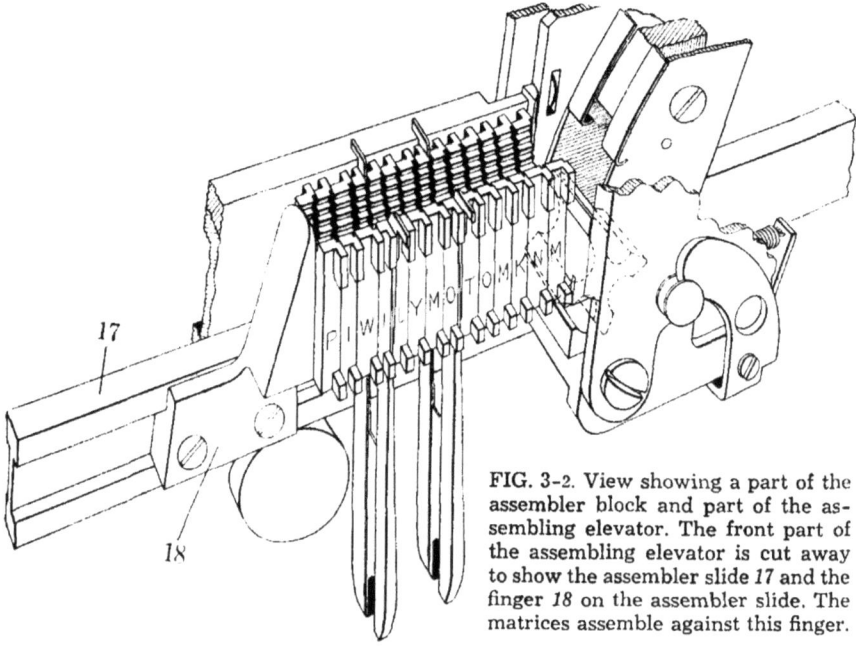

FIG. 3-2. View showing a part of the assembler block and part of the assembling elevator. The front part of the assembling elevator is cut away to show the assembler slide *17* and the finger *18* on the assembler slide. The matrices assemble against this finger.

Fig. 3.7 The Linotype *assembler block* containing a row of matrices. The two components protruding below the block are *spacebands*, which insert spaces between characters. Source: (Mergenthaler Linotype Corporation, 1940, Figure 3-2, page 40). Work is in the public domain in its country of origin and other countries and areas where the copyright term is the author's life plus 70 years or fewer.

in any case are arranged linearly. Parhami (2019) gives a nice introduction to printing for Persian and notes the introduction of Linotype machines in Iran in the 1950's.

However, the Brahmic scripts—Bangla, Gujarati, Tamil, Devanagari— would seem to present a problem, since the components of the script are not combined linearly: vowel diacritics in Devanagari, depending on the diacritic, may be written above or below the consonant group they pertain to. Consonants are joined together into often quite complicated ligatures. Thus the Sanskrit word मृत्यु ⟨mṛtyu⟩ 'death', is written in Devanagari with a subscripted ⟨ṛ⟩ (a syllabic /r̥/), त्य ⟨ty⟩, which is a ligature of त and य, and a subscripted ⟨u⟩. How can a Linotype machine, which assembles mats in a line, deal with this?

The trick, which was invented by Hari G. Govil, is to imagine slicing a line of Devanagari text into 'sectional characters' (Govil, 1937), and defining the granularity of those sections so that there are no more than 90 sectional types. Since Devanagari does not have an upper/lower-case distinction, this turns

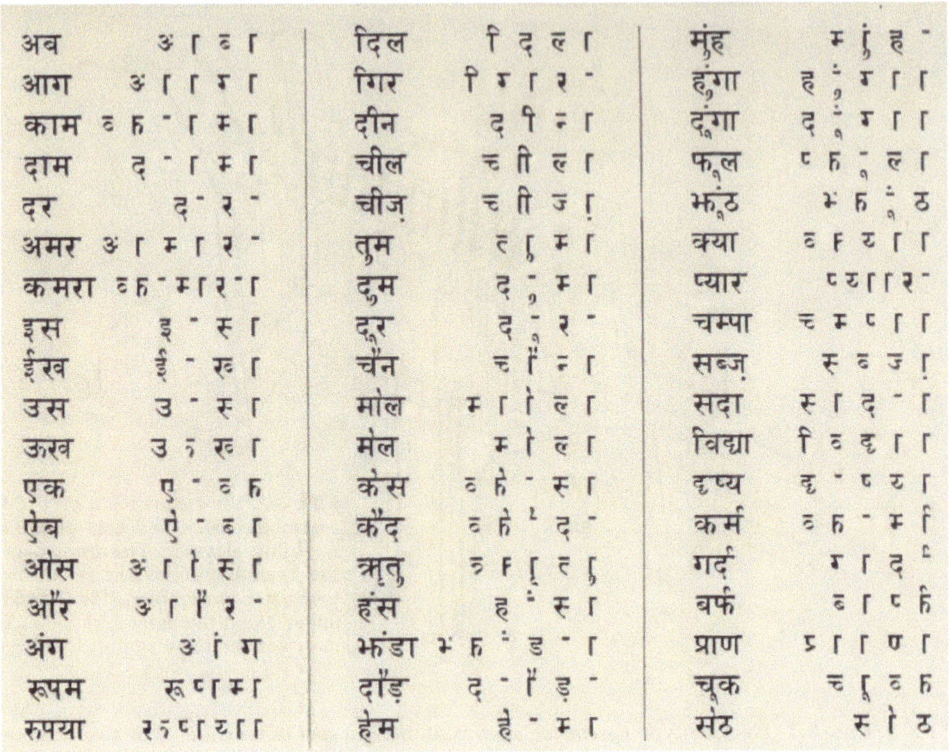

अब		दिल		मुंह	
आग		गिर		हुंगा	
काम		दीन		दुंगा	
दाम		चील		फूल	
दर		चीज़		भूंठ	
अमर		तुम		क्या	
कमरा		दुम		प्यार	
इस		दूर		चम्पा	
ईख		चैन		सब्ज़	
उस		मोल		सदा	
ऊख		मेल		विद्या	
एक		केस		दृष्य	
ऐख		कैंद		कर्म	
ओस		ऋतु		गर्द	
औरर		हंस		बर्फ	
अंग		भंडा		प्राण	
रूपम		दाँड़		चूक	
रुपया		हेम		संठ	

Fig. 3.8 Composition of words written in Devanagari using Linotype sectional characters. Source: (Mergenthaler Linotype Corporation, 1933, p. 7). Work is in the public domain in its country of origin and other countries and areas where the copyright term is the author's life plus 70 years or fewer.

out to be possible. Importantly, the sectional characters on the Devanagari Linotype machine are not always actual letters. For example the subscript ⟨u⟩ in the example दुम ⟨duma⟩ (seventh from the top, second column in Figure 3.8), is composed of the subscripted vowel in combination with the top bar.

The other Brahmic scripts supported by Linotype could be handled similarly, though there were difference of detail in how each script was handled. See Ross (1988); Oswald (2019); Ross (2021); and Malli (2022). Ross (1988), in particular presents a very detailed account of the development of printing technology for the Bangla script.

The resulting quality was, to be sure, suboptimal.[8] See Naik (1971); Ross (1988) for critical analysis. As we will see momentarily, Monotype allowed for somewhat better quality, but in the West, Linotype was preferred over Monotype in the newspaper industry since the full line casting made it possible to use the system in the rotary plates that were preferred by newspapers, thus allowing for more rapid printing. To some extent these considerations were less relevant in India where many newspapers could not afford rotary presses (Jeffrey, 2000, p. 37), and those that were introduced were for English-language newspapers. However, it remained true that Linotype was a somewhat cheaper process than Monotype, and the latter tended to be used more for book printing, where smaller output volume was needed, but higher print quality required. As Ross (2021) notes (p. 3):[9]

> The implementation of the Devanagari script on the Linotype composing machine, initiated in America by Hari G. Govil approaching the Mergenthaler Linotype Company (MLCo), was heralded for facilitating speedy newspaper composition (and thus dissemination) that would ideally, for the Indian vernacular press, match that of the 'English dailies' ...[T]he line casting system produced stable lines which could be more-easily be arranged into pages for stereotyping to make curved plates for rotary printing that was favoured by the newspaper industry. Yet, the Linotype was severely constrained by its 90-channel keyboard, which therefore only permitted a limited set of matrices in the main magazine. This limit of 90 characters per typeface could be augmented by a side-magazine of auxiliary characters, but the latter was expensive and slowed down the composing process—a factor antithetical to the exigencies of rapid newspaper production.

To understand why Monotype could produce higher quality than Linotype it is necessary to step back momentarily and look at hand-set movable type for Indian scripts, and in particular how diacritics were handled. There were two methods, illustrated in Figure 3.9. The first of these, the Degree system was more intricate in that it involved assembling type in both a horizontal and vertical direction, with diacritics assembled above or below the main letter (Figure 3.9, lefthand examples). The Akhand system took advantage of kerning, allowing adjacent type to overhang succeeding or preceding type, but requiring only horizontal assembly (Figure 3.9, righthand examples).

Unlike Linotype, Monotype, which cast single type blocks rather than whole lines, could implement kerning thus allowing for the use of the Akhand

---

[8] Over and above the worse type quality, there were also other problems introducing Linotype for use with native Indian scripts. Singh (2018, 2021) discusses the fact that, though Mergenthaler was an American company, its Indian operations based in Calcutta were subsidiary to the British branch of the company. This made it easy to perceive Linotype as being a colonial product that was being imposed on India. Over and above these political considerations, there was a mundane economic point. Linotype was a major advance in America because it allowed massive reductions of cost compared to expensive manual typesetting. But in India the situation was the reverse: mechanization was expensive, whereas the cheapest and most plentiful commodity around was labor.

[9] Page references to Ross are for the open online version rather than the version behind the journal's paywall.

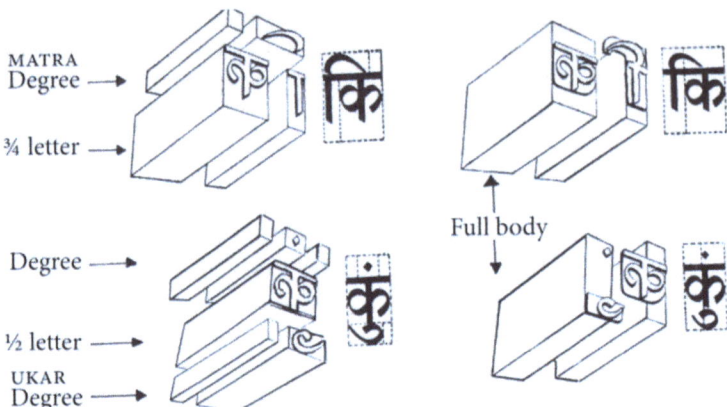

MATRA
Degree →

¾ letter →

Degree →

Full body

½ letter →

UKAR
Degree →

Fig. 3.9 The Degree (lefthand) and Akhand (righthand) systems of type composition, examples from Devanagari. Note in the example on the top right that the hook on the type for ि in the composition of कि is impossible for Linotype, but could be handled to some degree by the Monotype system. After Ross (2021, p. 113), originally from Naik (1971), Volume 2, p. 328, Fig. 96.

system. This facilitated the production of higher quality type and was preferred for book printing. Yet, even here, the quality was not up to that of hand-set type. Again, Ross (2021) (p. 2):

> The 'inauguration of the Devanagari Monotype machine,' the result of a commercial undertaking involving the exertions of key figures across continents, took place in Poona in 1932; while the Devanagari Linotype machine—the product of another trans-continental enterprise—was launched in Calcutta in 1933. The Monotype machine, developed by the British Monotype Corporation, which employed a form of the Akhand system was preferred over the Linotype for good quality work in the Devanagari script, yet it could not meet the exacting standards that Monotype set out:
>
> > "However ingenious a machine may be, whatever the speed it may attain in composition, quality in printing surface and facility of correction are essential features of its practical utility. No machine that is devised to do what previously was done by hand can be looked upon as correct in principle or perfect in adaptation, unless it maintains equal quality and greatly increased quantity of production. In such a degree as the machine lowers the standards created by the slow evolution of manual methods, it must be pronounced a failure."
>
> [cited in *The Monotype Recorder*, 39, no. 1, (London: The Monotype Corporation, 1949) p. 16.]

Indeed, one of the problems was that while Monotype could handle kerning, the overhanging diacritic signs were fragile and liable to breakage while printing (Naik, 1971, Volume II, p. 397).

While hot-metal printing technology had its drawbacks for Brahmic scripts, it was at least passable provided the printer and reader were willing to make some compromises. As noted above, Linotype or Monotype systems were developed for "most Indian scripts deemed wealthy enough to merit manufacturing the equipment" (Jeffrey, 2000, p. 36). Even the technically difficult Odia script, with its many vertically stacked conjunct consonants, and for which printing had long been a side show (Shaw, 1977), had a Linotype system designed for it (Jeffrey, 1997, 2000).

3.6.4.2 Hangul

In contrast to the situation in India, Linotype made no inroads in Korea. This is not to say that there were no Korean newspapers that used the technology. In fact as we shall see below there was one Korean-language newspaper that introduced a Linotype-derivative system as early as 1915. That newspaper, however, was based in California.

But before we describe that system it is worth stepping back and considering how Linotype would work for Korean. The first point that one must realize is that it makes a crucial difference which way the printing is done: Left-to-right, as with most modern Korean text, including newspapers. Or top-to-bottom, which was the traditional arrangement.

Left-to-right horizontal printing for Korean cannot support Linotype. For horizontal printing, if one were to apply a sectional character approach, one would need to cut the syllable combinations vertically. And the problem there is that there are too many syllable combinations and there is no way to use Govil's trick of cutting vertically to significantly reduce the number of needed type.

The Unicode code block for Korean contains over 11,000 Hangul syllable combinations, but this is an absurd overestimate of the number actually needed to print Korean text. The actual number of combinations in anything like common use is about 2,000. Lists of these can be found online.[10] Of course, even 2,000 is too many to fit on a Linotype keyboard with 90 keys, but more to the point, there is no way to reduce this number by cutting the syllables vertically. Consider the lefthand example in Figure 3.10 where we try cutting the syllables of the word 한글 *hangeul* into two components each. Pursuing this course, and asking roughly how many components one would need, we end up with a little over 1,900 types,[11] which affords hardly any reduction whatsoever.

But what if, instead, we print in the traditional way, that is vertically? For Linotype this would mean producing the slugs horizontally, with the syllables flipped 90 degrees, and then turning the slugs 90 degrees back when printing.

---

[10] For example at `http://nlp.kookmin.ac.kr/data/syldown.html`, retrieved June 1, 2024.

[11] See `https://github.com/rwsproat/hangul` for code.

Fig. 3.10 Two ways of cutting the word 한글 *hangeul* for a Linotype system. For horizontal left-to-right text one would need to cut vertically, but this would scarcely achieve any reduction in the number of needed type. On the other hand, with the traditional top-to-bottom direction, one can cut the syllable blocks horizontally, which can achieve a very large reduction in the number of needed elements. The righthand column shows how one can construct a new syllable out of the top element of 한 and the bottom element of 글. Source: own work.

In that case one is cutting the syllables horizontally (i.e. vertically, flipped 90 degrees), so that the syllables would be cut as in the central example in Figure 3.10. Using the top 하 of 한 and the bottom ㄹ of 글, one could then make 할. See the righthand exmaple in Figure 3.10. Done this way, the system becomes much more practical: one would need about 290 types in total, but in fact one could cover 99.99% of Korean text with only 196 types, which will fit on a somewhat expanded Linotype keyboard assuming one has two magazines.

This difference between horizontal and vertical sectionability relates directly to how Hangul was designed in the first place. In the 15th century, all texts in Korea were written vertically as they had been for over a thousand years since the introduction of Chinese writing into the Korean peninsula. So Hangul was designed with this direction in mind. At the macroscopic level, Hangul is linear (as is the case with any fully developed writing system), where the linearity runs vertically on the page. As discussed in Sproat (2000), and in Section 2.3.3 of this volume, the combination of elements in a script can deviate from the macroscopic direction, but only very locally— typically at the level of the syllable. Even within those units, the macroscopic direction is often the preferred one, so that for example in a Devanagari *akshara*, consonant sequences *mostly* involve left-to-right concatenation of the consonant glyphs.

But Hangul was designed to be written top-to-bottom, and within syllable blocks, half of the basic vowel symbols, and *all* of the final consonants, are written below (that is *after*) initial consonants. Thus even within syllables, most of the combinations follow the macroscopic direction. So while one must still spend a large proportion of the type (approximately 80%) covering the initial CV combinations where the vowel appears to the right of the consonant, the vowels written below, and the final consonants, can be separated off, affording a massive reduction in the number of needed type.

High quality type for Hangul has syllable blocks all the same size for a given sized font, so that the individual *jamo* must be resized and to some extent reshaped to fit. That of course is not possible with Linotype. But if—as with Kong Pyung Woo's Korean typewriter, which we will discuss in Section 3.7—one is willing to forgo esthetics for practicality, vertical Linotype typesetting is possible for Korean.

However, there has been only one case where this possibility was actually put into practice.[12] In 1909 the *Shinhan Minbo* (신한민보, 新韓民報, English title *The New Korea*) was founded in San Francisco with the goal of serving the Korean expat community, and as a voice for the Korean independence movement. In 1915, Rev. David Lee (리 대위, modern spelling 이 대위), who subsequently became the editor of the paper, invented a typesetting system based on the Intertype system. Intertype used essentially the same technology as Linotype, and was in fact founded by former Mergenthaler employees after the original Mergenthaler patents expired (Bell, 2011). Lee's system started with a single keyboard to cover the most common combinations, with side sorts for characters that fell outside that set. This was later changed to a system with two magazines, which could cover 196 types. Article titles were still hand set (and often involved Chinese characters), so that there was a marked difference in quality between the titles and the main body of the articles.

It should be pointed out that the publications using David Lee's system used an old spelling system that predated several reforms that were introduced in Korean orthography at various points in the 20th century.[13] The older spellings were in some ways simpler than the newer spellings, so that for example there were fewer consonants that could appear in the final position (bottom) of the syllable, and no consonant sequences.[14]

The New Korea March 11, 1915 edition published a story about Lee and his new typesetter, and the April 15, 1917 edition published the complete

---

[12] See Bell (2011) for in-depth discussion of various attempts to introduce linecast and monotype printing for Korean. We are also grateful to Aaron Bell for discussion about these issues.

[13] See https://namu.wiki/w/%ED%95%9C%EA%B5%AD%EC%96%B4/%EB%A7%9E%EC%B6%A4%EB%B2%95/%EC%97%AD%EC%82%AC for discussion of Korean spelling reforms.

[14] The calculations discussed above for the needed types are based on modern spelling, and therefore probably yield somewhat of an overestimate for the number of types needed.

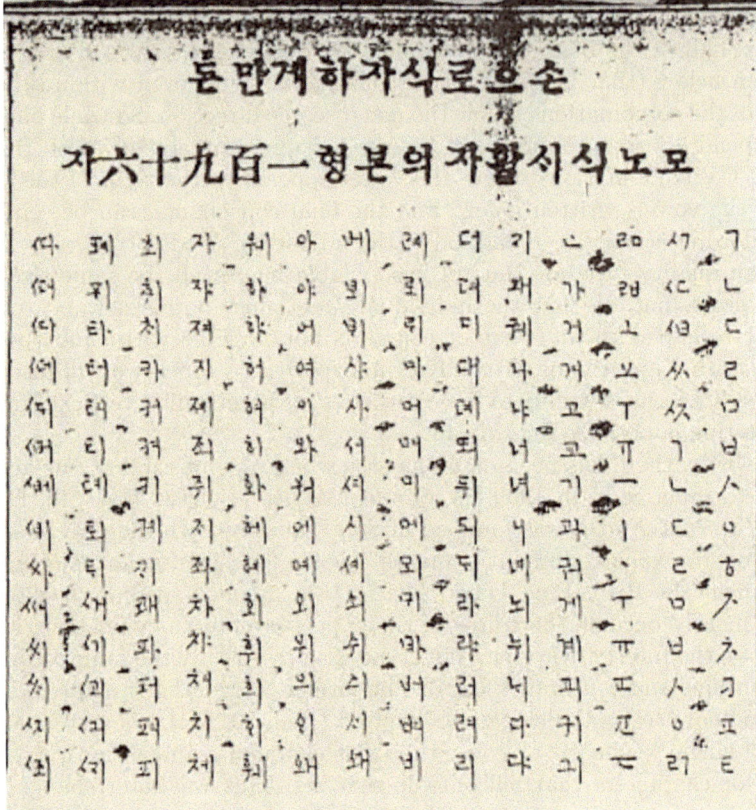

Fig. 3.11 The complete set of type from *The New Korea*, April 15, 1917. Work is in the public domain in its country of origin and other countries and areas where the copyright term is the author's life plus 70 years or fewer.

set of type on its front page (Figure 3.11). Not every type that was needed was covered, and occasionally it was necessary to cut corners. For example, 께 ⟨kke⟩, which in the old spelling appeared as 쎄 ⟨ske⟩ was apparently missing from his set, so he had to print this with a ㅅ ⟨s⟩ above 게 ⟨ge⟩.

Lee's Intertype system appears to have been the only successful deployment of a sectional character hot-metal typesetter for Korean newspaper publishing. Other machines of the same design appear to have been ordered for use: for example the March 21, 1946 English edition of *The New Korea* reports that three Intertype machines were ordered for Korean publications, all outside Korea. The same article also reports that apart from *The New Korea*, another publication, *Korean Independence* was also using the same machine. But the technology appears to have made no inroads in Korea itself, even after independence from Japan. Bell (2011) documents various attempts by the Linotype and Monotype corporations to work with Korean newspaper publishers, and while some of these produced designs, none of them apparently made it to production. On the other hand, 'Monotype' machines of Japanese designs, with large matrices like the one discussed in Section 3.6.2, were well adapted for Korean (the 2,000 commonly used Korean syllable blocks could easily fit on such devices) and continued to be used.[15]

3.7 Typewriters and script complexity

Typewriters have been a popular topic of late, with a great many books that document their 100 years of use. Yet few of these works even mention, let alone discuss in any detail, the adaptation of typewriter technology to non-linear scripts, or scripts with very large character sets. The main exception to this is the work of Mullaney (2012, 2017, 2021) on Chinese typewriters, and the similar Japanese typewriters. What makes Chinese interesting—and a *reductio ad absurdum* of the point—is the character set size, which makes it intractable for the technology we normally think of when we think of type-writers. But between the two extremes of English and Chinese, are many scripts for which typewriters of the more conventional sort were developed, but where the development was not simply a matter of replacing the keys and the strikeplates with a different character set. Again, these scripts have scarcely been discussed at all in the literature, so we hope that our discussion below will help fill this void.

We defer until later (Section 5.3) a discussion of the history of typewriting since much of that discussion relates to the initial designs of the keyboard layout as they pertained to gesture complexity. Here we focus on how the technology of typewriting was adapted beyond their original applications to linear scripts.

3.7.1 Hangul typewriters

An important difference between printing technology and typewriters now comes into play. In printing, one sets a whole text—a whole page of a book, or a column in a newspaper—inks the set type, and prints. In typewriting, the individual characters are transferred to the page sequentially. For scripts that combine characters in a non-linear fashion, this means that there is in principle a way to construct a composite glyph out of component glyphs by

---

[15] Bell (2011), page 121, quotes from the report by Alan Barke of the Monotype Corporation (1969), that "Japanese automatic type casting machines were already setting 'Hangul' with the addition of Chinese characters and in fact his company has just installed six Keyboards and three Casters manufactured by Ikegai, Tokyo", Ikegai being one of five 'Monotype' style machine manufacturers mentioned in CONPT (ND).

imprinting each component glyph in a sequence, assuming one can provide the typewriter with a mechanism whereby it will not always advance the carriage after each keystroke.

The upshot of this is that whereas Linotype technology was impossible for Korean if one wanted to print the text horizontally, it was in fact possible to build Korean typewriters, though some ingenuity was required in their design, and the result was still not as pleasing as handwritten text, or modern computer fonts. We already saw an example of Korean typewritten text in the Korean Armstice document (Figure 1.3), and commented on some of the esthetic issues.

Korean typewriters have a fairly long history, with the first model being developed in approximately 1914. But the development really took off after 1945, with the end of World War II and the independence of Korea from the Japanese Empire. Under the Japanese imperial occupation, the use of Hangul was not encouraged, with the occupiers favoring a Japanese-based education, with the goal of turning Korea into a nation of second-class Japanese citizens. The Hangul script, which had long been a source of national pride in Korea, took on an even greater role as a symbol of nationalism. The earliest newspaper to use Hangul exclusively was the *Tongnip Sinmun*, published 1896–1899 during the pre-colonial period during the last days of the the Joseon dynasty. As the name ('Independence Newspaper') implies the paper was outwardly nationalistic, and countered the increasing foreign interference in Korean national affairs. Later overseas Korean newspapers, such as the *Shinhan Minbo*, which we discussed above, were uniformly pro-independence publications, and uniformly used Hangul in their editions—though even *Shinhan Minbo* used Chinese numbers, and also used occasional Chinese characters in story titles. In contrast, newspapers published on the Korean peninsula during the Japanese occupation, and even for several decades afterwards, used a large number of Chinese characters (*Hanja*).

After independence, however, the move towards eliminating Chinese characters from everyday use gained momentum. This was complemented by moves towards script reform, westernization and the increased adoption of modern technology. All of these were seen as crucial for bringing Korea out of its traditional cultural dependence on China, followed by thirty-five years as a colony of the Japanese Empire.

Part of the move to technologize Korean involved typewriters, and one of the points that became apparent quickly was that Korean writing was not trivially adaptable to the new technology. The traditional vertical writing direction which proved to be essential to the adaptation of linecasting hot-metal printing technology, was problematic for typewriters, which were geared to a horizontal orientation—at least if one wanted to easily read what one was typing.

In fact, the 1914 typewriter developed by Lee Won-ik, a Korean-American living in Hawaii, based on the American Smith Premier No. 10, did produce vertical text, but one had to type the vertical line horizontally, and then flip the paper at the end to read it (Kim, 2023). Lee's machine, which had 84 keys arranged in 7 rows, used what would later be known as a five-set keyboard, with two sets of initial consonants, two sets of vowels, and one set of final consonant, as well as keys for Chinese numbers. Onset consonants and a subset of the vowels were associated with *dead keys* so that the carriage would not move, but to get complex (emphatic) initial consonants, as in 께 ⟨ske⟩ (old spelling for 께 ⟨kke⟩), one had to rotate the platen manually in order to position the two consonants horizontally next to each other (vertically next to each other at typing time). But again this approach had the drawback that one would have to crane one's neck at 90 degrees if one wanted to read what one was in the process of typing. Besides vertical text, another problem was Chinese characters, assuming one wanted to include them. Indeed the only solution to that was to adopt a Chinese-typewriter-style approach (Section 3.7.4) and indeed such systems did exist.

Both of these problems could relatively easily be dispensed with if one agreed only to use Hangul, and to type text from left-to-right rather than top-to-bottom, and indeed all of the discussion surrounding the best approach for Korean typewriting at least agreed on these two points. And both of these— linearization from left-to-right, and the exclusive use of Hangul—fit well with the goals of the post-liberation Korean modernizers. It also fit well with the new world order that included US domination of much of the Western Pacific. The US military, as well as the US-educated first President of the Republic of Korea, Syngman Rhee supported westernization in general and the use and mechanization of Hangul in particular. October 1, 1948 saw the enactment of a law mandating the exclusive use of Hangul in official documents.

Still, the Hangul script was tricky for typewriting. While David Lee's Intertype system allowed for coverage of most Korean text with 196 types, this number was not practical for typewriters, without significantly expanding the keyboard. The number of basic letters (*jamo*) is small, and could easily fit onto a typewriter keyboard: the problem was how to combine them. There was of course one simple approach: dispense entirely with the traditional grouping of *jamo* into syllable blocks, opting instead for a total linearization of the script. This idea had actually been around for some time. David Lee's *Shinhan Minbo* printed its masthead this way for a few years during the 1920's (Figure 3.12), though none of the columns in the paper appear to have been printed that way. Linearization had some proponents, including the Joseon Language Research Society (predecessor of the present-day Hangul Society), and various scholars such as the Korean language scholar and linguist Choi Hyun-bae. For Choi, linearization was more logical in that, in his view, it aligned with the principles of Korean phonetics, and followed a "natural" system used by many writing systems worldwide. Indeed, Choi went on to propose a linearized version of Hangul, that was somewhat modified in form so as to make the letters look more pleasing when strung out in a line (Bell, 2011, page 51). More to the point, it would simplify the script and facilitate

Fig. 3.12 Complete linearization of the newspaper's name (신한민보) in the masthead of the *Shinhan Minbo*, January 22, 1924. This design ran for a few years in the early 1920's. Work is in the public domain in its country of origin and other countries and areas where the copyright term is the author's life plus 70 years or fewer.

mechanization: indeed a Korean typewriter built on these principles would have been straightforwardly adaptable from a standard English typewriter.

Linearization was, however, deemed too radical a change, and in 1954, Rhee's government decreed "assembled writing", i.e. the traditional syllable-block grouping of *jamo*, as the standard, which effectively brought to an end proposals for linearization.[16,17]

Once "assembled" writing was agreed upon, the issue became how to produce reasonable-looking text as easily as possible. In practice the two considerations—quality of output versus ease of input—had to be balanced against each other. The easiest systems in terms of the effort of the typist were three-set systems that separated the letters into initial (onset) consonants, final (coda) consonants, and vowels. Such systems necessarily compromised esthetics heavily in favor of efficiency, but were nonetheless popular for this latter reason. The best known of the three-set systems was due to Kong Pyung Woo, who took out a US Patent on his design in 1953 (Kong, 1953). Kong himself was an interesting character. Trained as a doctor in Korea during the Japanese imperial period, Kong became Korea's first ophthalmologist, opening a clinic in Seoul (1937) that is still in operation today. One of Kong's

---

[16] That said, the linearization of Hangul for technological purposes resurfaced in the 1961 proposal by the Mergenthaler corporation for a linearized Hangul Linotype system: see Bell (2011) for discussion. But as Bell makes clear, many of the American printing technology corporations who hoped to enter the Korean market really had little clue as to the technical, cultural and political forces they were up against, and Mergenthaler's proposal never made it past the draft phase.

[17] One other argument against linearization is that it had the potential to make reading slower. There is of course no meaningful comparison that can be done between traditional "assembled" writing in Hangul, and linearized versions, since nobody is fluent in reading the latter. But for people who can read Japanese, there is a definite sense that reading Kanji is faster than reading Katakana, for the simple reason that the former take up less space in the visual field. This is particularly noticeable when one is presented with text in streaming mode, as for example in streaming text LED displays on trains. There is some psycholinguistic support for Katakana being in general slower to read (Osaka, 1989), though see Jincho et al. (2014) for some criticisms of Osaka's experimental design. In any case it is clear that linearized Hangul would take up more horizontal space in the visual field than "assembled" Hangul, which could in principle lead to slower reading, not to mention the fact that more paper would be needed.

patients, a Korean-language specialist, complained of problems with writing in Korean. This inspired Kong's interest in typewriters, eventually leading him to found the Kong Pyung Woo Typewriter Company. According to the text of Kong's US patent, Kong's goal was to "provide an improved Korean language typewriter, which will remedy [defects with previous Korean typewriter designs] ...[and] which will possess all of the aforesaid advantages of the invention, and utilize substantially all and identical parts, features, and operations of the conventional American machine, whereby the typewriter may be manufactured with ease, convenience, economy, and resort to the already existing facilities of the manufacturer of such machines, and hence resulting in low purchase price to the consumer."

In Kong's design, the *jamo* were divided into into those that, when the corresponding key was struck, the carriage would move to the left; and those where striking the key did not cause the carriage to move—i.e. dead keys. The former were placed on the righthand side of the keyboard, and consisted largely of the syllable-initial consonants, as well as symbols where one would typically expect a carriage advance: punctuation, numbers, and the *won* (currency) symbol. The dead keys were on the lefthand side of the keyboard and consisted of the vowels and final consonants.[18] See Figure 3.13. Note the characters on the bottom and top of each key, controlled by the shift key as on an English typewriter. The movement of the carriage was controlled by having the keys pass through one of two slots on its way to striking the platen. The righthand keys would pass through the righthand of the two slots, which would actuate a gear that would cause the carriage to move to the left. The lefthand keys would pass through the lefthand slot, which did not activate this gear.

Kong's patent describes how to type the syllable 왔 ⟨wass⟩: first the initial ㅇ from the righthand side of the keyboard is typed, causing the carriage to advance. Then in sequence are typed ㅏ , ㅜ and ㅆ . Since these are on the lefthand of the keyboard they do not advance the carriage. Also—and this is another clever feature of Kong's design—since these keys all strike the platen through the lefthand slot as described above, the upshot is the characters show up aligned with the initial ㅇ , even though the carriage has already advanced. To start the next syllable, the typist would then type the space bar, which also advances the carriage as on an English-language typewriter. Figure 3.14 shows an example phrase containing the syllable 왔 as typed on Kong's machine.

A couple of observations are in order. First, because of the division into initials and finals, there is duplication on the keyboard. Thus ㅇ , for example, can be found on both the righthand and lefthand portions of the keyboard, corresponding to its use as an initial and final, respectively. The initial ㅇ is labeled as '26a' in Figure 3.13; the final is on the leftmost key on the second

---

[18] The one exception to this pattern was the very rare final consonant sequence ㄹ ㅍ ⟨lp⟩, which was placed on the upper righthand key, presumably because Kong ran out of space to put it on the lefthand side.

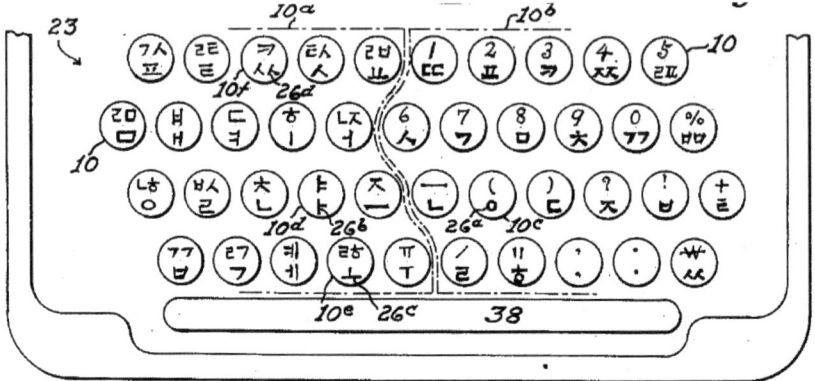

Fig. 3.13 The keyboard of Kong Pyung Woo's typewriter, from Kong (1953). The keys to the left of the meandering line in the middle of the keyboard consist of vowels and final consonants, and do not advance the carriage when struck. Those to the right are initial consonants, which cause the carriage to advance, along with other symbols such as punctuation and numbers, where one would generally expect a carriage advance. Source: US Patent.

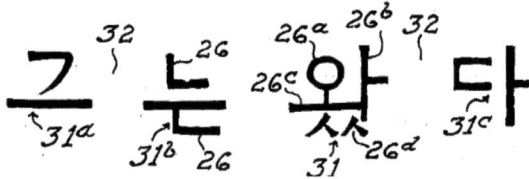

Fig. 3.14 The phrase 그는 왔다 *geuneun wassda* 'he came', as presented in Kong's patent document (Kong, 1953). Actual output from Kong's typewriters did not look quite this nice. Source: US Patent.

row from the bottom. This division into three basic character sets—initials, medials and finals—is termed the *three-set* (세벌식) keyboard, as noted above.

Secondly, the design of course did not allow for the various sizes of characters that one finds in well-balanced fonts (see Figure 3.15 below), but it did not even allow for simple allography such as that found in ㄱ ⟨g⟩, which changes its shape depending on whether the following vowel is to its right or below it. In Kong's machine, the form of ㄱ used for 그 ⟨geu⟩ is the form that would be appropriate for 기 ⟨gi⟩. Presumably this was because Kong ran out of space on the keyboard, or adding these variants would have complicated the typist's task, or because the allography was not considered to be that important.

But in fact, the problems with the esthetics went well beyond this. Another problem with Kong's design was the 'clothesline font' (빨랫줄 글꼴) property whereby the height of the 일 ⟨il⟩ was greater than that of 이 ⟨i⟩. On a Kong

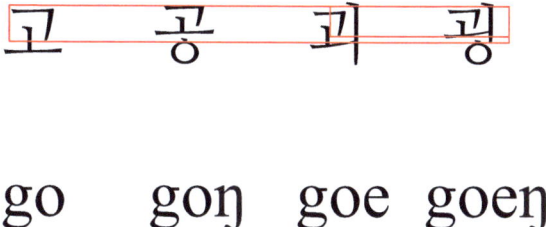

Fig. 3.15 Four instances of initial ㄱ ⟨g⟩ in syllables of different complexities in a modern computer font. The initial consonant gets progressively smaller as the complexity of the syllable increases. Source: own work.

typewriter, there was no difference between the form of 이 as an independent syllable and when it occurred as part of 일, so that the final ㄹ seemed to 'hang down' from the syllable. This can be seen in the the Korean Armistice document in Figure 1.3.[19]

To get away from these problems, and render Hangul syllables in a fashion more akin to what one found in handwriting, traditional printing, or modern computer fonts, one needed to divide the sets further. Indeed there were four-set and even five-set designs. The four-set design distinguished between vowels that were also final in their syllable (이), versus those that were medial (일). This would solve the problem of text alteration noted in footnote 19, but it was still not enough to make the text as esthetically pleasing as one might like. For not only do vowels change their dimensions depending on what follows, but even initial consonants change their size depending on the complexity of the syllable. See Figure 3.15. This latter point was addressed, but only partially, by the five-set arrangement, which allowed for more than one variant for initial consonants.

The four- and five-set machines had an obvious disadvantage compared to Kong's design in that they were much harder to type on. Thus Kong's three-set typewriter can be seen as a compromise between what is desirable, and what is practical or efficient in producing text in a non-linear script, similar in spirit to the compromises needed for Linotype for Hangul and Brahmic scripts that we discussed above.

---

[19] See https://namu.wiki/w/%ED%83%80%EC%9E%90%EA%B8%B0, where it is also noted that one of the weaknesses of the design was that it makes it easy to alter a text: 이 can mean 'two' in Korean and one of the meanings of 일 is 'one', so one could easily change 'two' into 'one' by adding a ㄹ at the bottom, and this would not be noticeable.

Kong's system was popular initially with the military (which presumably explains its use to type the Armistice document), as well as in other government organizations and in commercial companies.[20]

But esthetic considerations remained an important factor. Five-set systems were harder to use for the typist, but produced better-looking output, and one particular design by Kim Dong-hoon (1957) became prominent in the 1960's. However it was ultimately a government mandate for still another four-set system that set the standard, particularly as this became mandated for use by various government ministries (Kim, 2023).

Hangul typewriters, despite their complexities and suboptimal output, must be considered a success. If nothing else, typewriters became commonplace in Korea, and they were not limited to government offices or private companies; even some ordinary families owned them. It is noteworthy that neither of the two prominent inventors of Hangul typewriters—Kong Pyung Woo and Kim Dong-hoon—were trained typewriter engineers. Like David Lee, they were motivated purely by the practical desire of mechanizing Korean text input. But these efforts happened in tandem with script engineering—including deprecating the use of Chinese characters, and switching from a top-to-bottom to left-to-right writing direction—which the government implemented in consultation with language experts (Kim, 2023).

3.7.2 Typewriters for Brahmic scripts

Typewriters were also developed for Indian Brahmic scripts, with the first Devanagari typewriter, built by Remington, introduced around 1930.[21] Indian-government-approved layouts for Devanagari keyboards were refined in the 1960's. Also in the 1960's, the India-based Godrej company introduced a line of typewriters including designs for various Indian scripts, which proved to be very popular.

In Devanagari, most consonants have a 'full' and 'half-letter' form, the latter being used when the consonant is ligatured with a following consonant. Thus त ⟨t⟩ when combined with a following क ⟨k⟩, loses its vertical bar so that the composite looks as follows: त्क. For the most part the keyboard included the half-letter forms for the consonants, with the vertical bar (*halant*) having its own key. If one wanted to type त, one could type two keys, one for the

---

[20] Unfortunately, though, Kong was not successful in one of his main goals, namely manufacturing the machine at a price that could make it a consumer item. A 1965 advertisement for the "Prince 5" model promotes it as Korean, rather than American-made, and at a price of 29,900 won, which in those days would have been equivalent to several months' of an office worker's salary. See `https://namu.wiki/w/%ED%83%80%EC%9E%90%EA%B8%B0`.

[21] See Sinha (2009), for a discussion of Devanagari typewriting, from which much of the material in this discussion is drawn.

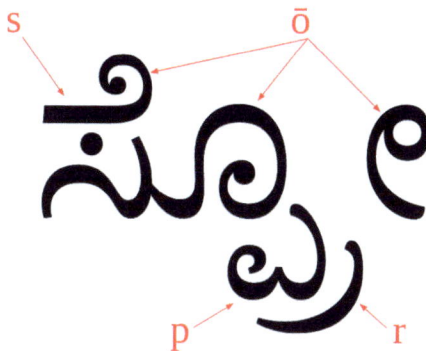

Fig. 3.16 The syllable ⟨sprō⟩ in Kannada script. Note the first consonant ⟨s⟩ written inline, with the next two consonants written as subscripts. The vowel is written as a combination of three components, including a diacritic on the initial consonant, an inline component ligatured with the initial consonant, and a second inline component on the righthand edge of the syllable. Source: own work.

⟨t⟩ and one for the vertical bar. However, space on the keyboard permitting, composed full-letter forms were included for the more common consonants. Vowel diacritics that needed to be placed above or below the consonant, were included on dead keys that did not move the carriage, or on 'half-backspace' keys that moved the carriage back—the latter similar to the way a ⟨ç⟩ could be typed on an English typewriter by typing a ⟨c⟩, then backspacing and typing a comma. More complicated combinations, such as stacked consonants—e.g. द्द for द+द—could be obtained by rotating the platen. The design of the Devanagari typewriter thus bore some similarity to the design of Devanagari Linotype in its approach to decomposing characters, the main difference being that print quality was not the goal, so the decompositions of characters could afford to be less draconian.

Using a Devanagari typewriter was not straightforward, and much skill was required to produce Devanagari text that looked plausible.[22] Still, as Sinha notes (page 13), "the quality of the typewritten Hindi text [was] poor, with broken lines, broken characters, and bad alignment." People used typewriters since they did save labor, and produced output that was at least acceptable. Again, the script forced a compromise between what was desirable and what could be technologically achieved.

---

[22] Raiomond Doctor, who worked on early computerization of Indian languages, reports (p.c.) that when computer-based input systems were first introduced, typists protested since the new input systems made complex tasks like creating conjunct consonants too easy. They demanded keyboards more akin to what they were used to on typewriters be developed so that they could still make use of the skills that they had learned.

Fig. 3.17 Keyboard of a Godrej Malayalam typewriter, redrawn from Manohar & Thottingal (2018), Figure 7a.

Fig. 3.18 Traditional Malayalam consonant conjunct combinations beginning with ⟨k⟩, with various vowels.

The situation was similar for other Brahmic scripts of India. The first Kannada typewriter was demonstrated in Mysore in 1940.[23] Later models, such as those of Godrej, were only slightly different from the original design. Unlike Devanagari, where consonant sequences involve ligatures of half-letter forms as described above, in Kannada (and also Telugu) script, the first consonant in a sequence is written inline, and the remainder are written as subscripts on the initial consonant (Figure 3.16). This could often be achieved by turning the platen a half turn. As with Devanagari, typing in Kannada was not very straightforward.

---

[23] We thank Professor KP Rao, Manipal Institute of Technology, for information about Kannada typewriters.

Similarly for Malayalam, while typewriters were common (Figure 3.17), they had many limitations when it came to producing esthetically pleasing output. The traditional Malayalam abugida had over 200 conjoined consonant glyphs (Figure 3.18), with many that were not readily predictable given their component consonants. These also had to be ligatured with vowel glyphs. Obviously it was impossible to fit these onto a typewriter, and one could not even design a typewriter that could build up the conjuncts from component pieces. So as a result, Malayalam typewriters kept only the most common conjuncts on their own keys, and for the rest the typist had to be satisfied with typing a linear sequence of consonants. As Manohar & Thottingal (2018) note (page 334), the resulting text was "a painful experience for reading and did not do any justice to the beauty" of the script. As was the case for Hangul, the conflict between the script and the technology became an issue of script engineering, but unlike the case in Korea, in the case of Malayalam, as Manohar & Thottingal (2018) (*ibid*) explain, the technology won the day (see also Mohanan (1996), page 422):

> To solve this problem, either the typewriter, or the language had to be redesigned. There were demands from newspaper and publishing industries to reduce the script complexity so that Malayalam becomes better suited for typewriters and printing. Based on this, in 1967 Kerala government appointed a committee to study script reformation. The committee submitted their report and in 1971 Kerala government published an order to reduce the complexity of the script ("Malayalam Script. Adoption of New Script for Use. Orders Issued" 1971).[24]

The difference between Malayalam and Hangul ultimately comes down to script complexity. Hangul was tricky to implement on typewriters, but not impossible if one was willing to compromise a bit on the look or put up with a bit of extra complexity during the typing process. As a result, the pressure to change the script was not sufficiently strong to overcome tradition. In the case of Malayalam, there was a clear disconnect between the script complexity and what modern (at the time) technology could handle, and so the script had to change. Ironically, a few more decades, and the advent of computer typesetting, made that change unnecessary.

Beyond India, typewriters were developed for numerous other Brahmic scripts such as Sinhala,[25] Thai, Burmese (Fowle, 2017), and even Tibetan with its complex stacked letters. Interestingly, movable type was never developed for Tibetan (Scherrer-Schaub, 2016; Ważny, 2016)—the technology there was woodblock printing—in large measure because of the complexities of the script. But, with some compromises, typewriters were developed: Parsons (2018) provides an interesting account of the development of the very first Tibetan typewriter during the 1950's.

---

[24] We note in passing that while the quotation refers to the "language" being redesigned, the issue obviously was not with the Malayalam language per se, but rather with its script.

[25] https://lankamarket.lk/ad/olympia-typewriter-sinhala/, accessed October 6, 2024.

Fig. 3.19 A 12pt IBM Selectric™ ball for Icelandic, purchased in 1980. The þ is in the middle at the bottom. Source: own work.

Like Hangul typewriters, typewriters for Brahmic scripts must be considered a success. They often produced quite poor quality output, necessitated by many compromises between the technology and the complexities of the scripts. But in India they were widely adopted in, for example, government industries, and there were many typing schools (since converted to computer schools) that trained would-be civil servants in this useful skill. Typewriters did not become household items in India, though this was presumably largely for economic reasons more than anything to do with the technology itself. This answers one of the questions posed in our introductory chapter, namely: if Jack Kerouac had been a Kannada speaker would he have typed? Not likely.

3.7.3 The IBM Selectric

The IBM Selectric™ was first introduced in 1961. Previous electric typewriters were essentially powered versions of their mechanical antecedents, that included the same strike-plate technology. The Selectric's innovation was to replace the complete set of strike plates with a single ball mounted on a shaft containing all the characters. Depressing a key on the keyboard activated the ball, spun it to the appropriate position and angle and caused it to hit the ribbon, thus printing a character on the paper as with the older strike-plate technology. The ball then moved to the next position—on the Selectric, unlike on traditional typewriters, the carriage did not move. The ball was interchangeable with other balls, meaning that one could switch between typefaces, font sizes, or between scripts, by simply replacing the ball. Figure 3.19 shows a Selectric ball for Icelandic, purchased by the second author in 1980.

The Selectric has been billed as the most successful typewriter ever manufactured, and indeed it was practically ubiquitous in offices up through the

1980's when computers started to take over. The ball could hold 88 characters, later raised to 96 with the Selectric III. It made switching typefaces trivial, and was easily adaptable to a large number of languages: linear alphabetic scripts were no problem, and right-to-left scripts such as Hebrew and Arabic[26] could readily be accommodated by simply reversing the direction of the ball's progress.

On the other hand, it did nothing to make it simpler to type in Hangul or in Indian Brahmic scripts. Indeed, as far as we are aware, there was no official adaptation of the Selectric to any of the Brahmic scripts of India. There were various adaptations for Hangul,[27] and the Korean patent disclosed in Kim (1978) describes one design for a four-set keyboard arrangement that allows for the setting of a larger number of dead keys, where the ball would not move, allowing for the stacking of *jamo*. But the resulting type quality was scarcely an improvement over previous Korean typewriters. See Figure 3.20 for an example of the output quality.

3.7.4 The Chinese 'typewriter'

Traditional typewriting technology was able, after a fashion, to cover scripts that combined elements in a non-linear manner. Typewriters existed for both Hangul and various Brahmic scripts, and while the results were not always pretty, they did at least work. But the technology is still limited by the number of keys and the practical levels of control for non-linear combinations. These limitations made it impossible to design typewriters for Chinese and Japanese.

True, machines that were called typewriters for these writing systems did exist. Mullaney (2017) presents a fascinating history of these devices, with the first part of his book devoted to debunking popular myths about them. A stock joke conception of the Chinese typewriter was a device modeled on an standard American typewriter, but with thousands of keys. Such devices never existed—though the Japanese 'Monotype' printing keyboard illustrated in Figure 3.5 does come close. Mullaney discusses various attempts to decompose Chinese characters into manageable sets so that one could type components of the characters to be assembled into whole characters in a fashion similar to the composition of Hangul syllables discussed in the previous section. But none of these attempts were successful.

---

[26] See, e.g., `https://www.phillytypewriter.com/arabic-selectric.html/` (accessed August 24, 2024).

[27] J. G. Savard's discussion at `http://www.quadibloc.com/comp/kyb0201.htm`, and a video demonstrating one of the models can be viewed at `https://www.youtube.com/watch?v=dMdYPaZ9Gmg&ab_channel=VintageTypewriterService` (both accessed August 24, 2024).

이 타자기는 세계적으로 알려진 아이비엠 셀렉트릭 타자기에 맞추어 개발한
한글볼로 타임핑 하였습니다.

한글은 세종 대왕께서 훈민정음을 개발하시고 수많은 한글 선구자분들에 의해
연구 되고 발전되어 오늘에 이르렀습니다.

Fig. 3.20 An example of text written using a Korean version of the IBM Selectric. Note the 'clothesline font' appearance of the text. Image provided by An Byeongjo (안병조) of Hello Kor Typewriter, used with permission.

What one had instead were machines along the lines of the one depicted in Figure 3.21. The machine had a matrix containing thousands of type—the Double Pigeon model depicted had 2,500 characters. The strike mechanism was shaped similar to a typewriter's strike key, but which ended not in a piece of type, but rather with a *holder* for the type. The operator moved the strike key to the desired character on the board. On depressing a lever, the strike key would pick up the desired type, imprint it on the platen, and then immediately return it to the matrix. The device was thus similar in spirit to a Monotype machine, though controlled by a different mechanism, and of course without the hot metal technology.

Fig. 3.21 A Double Pigeon (双鸽) brand Chinese typewriter at the Institut für Sinologie – LMU München. The case had room for for 2,500 characters. A full set of 7,000 characters was available, which could be switched into case as desired by the operator. Source: `https://en.m.wikipedia.org/wiki/File:Chinese_typewriter.jpg`, Author: Dadiolli/Tilman Schalmey, CC-BY-SA 3.0.

Even more similar in spirit to the (Japanese version of) Monotype were electric models with revolving drums embossed with characters, which could be positioned via a keyboard to impress the desired character on the page.[28] The operator had to press more than one key in order to select the desired character, and had to memorize the key combinations needed to produce thousands of characters. A good operator on an electric machine could achieve about 45 words per minute (Mullaney, 2021).

But despite the name, Chinese typewriters such as the Double Pigeon model were really machines for semi-automating the process of type selection—essentially replacing the hand of the traditional typesetter with a mechanical selection device. Electric models with revolving drums were indeed more typewriter-like, with some similarity to the IBM Selectric in particular, with the major difference that the selection of the character to be typed required multiple key presses, and substantial memory skills on the part of the operator.

---

[28] Earlier mechanical models with a drum also existed: the typist revolved the drum and used a pointer to select the desired character, which would cause the machine to select that character and print it.

Both because of their bulk, expense, and the skill required to operate them, Chinese typewriters never became household items. From both a technical point of view and the practical point of view of the needed skill to use them, Chinese 'typewriters' had more in common with mechanised printing technology than with typewriters.

3.7.5 Summary: The limitations of typing technology

Typewriters were never intended to produce the same quality output as a professional printing system. Even for English, the language and writing system best supported by typewriter technology, the difference in quality between a printed page and a typewritten page was obvious, even to the casual observer, and even with the best typewriters. Beyond this general point, though, there were few limitations imposed by a script with a small character set where the characters are arranged linearly, and do not need to be joined to one another.

Other scripts were not so lucky, and various compromises had to be made, balancing what was esthetically desirable in the output, with what was technically feasible with the machine, and practical for the typist. As we saw particularly in the case of Hangul, ingenuity was frequently needed to achieve this compromise, but the results were never entirely satisfactory. Further developments in typewriter technology, such as the IBM Selectric, while providing a revolution in typing for linear alphabetic scripts, did little to improve typing for Korean, and apparently left other scripts out in the cold.

Still, the odd compromises could leave a nostalgia-worthy legacy in some cases. For Korean, Kong Pyung Woo's 'clothesline font' has become somewhat trendy for those wanting a 'retro' look: For example it is used in the logo for the 과학동아 (*Donga Science*) monthly science magazine.[29]

3.7.6 Epilogue on typing: Beyond script complexity

We have focused in this section on how typing technology interacts with script complexity and how for many scripts typing has involved various compromises.

But the differences between typing and handwriting go well beyond difficulties with scripts. There is also an effect of the different input media on cognitive function. Van der Weel & Van der Meer (2024) report on an electroencephalogram study where participants (students at a Norwegian university) were required to copy words presented on a screen, either by handwriting with a digital pen on the screen, or typing the word with the right index finger

---

[29] See, e.g., `https://en.namu.wiki/w/%EA%B3%BC%ED%95%99%EB%8F%99%EC%95%84`, accessed May 3, 2025.

on a keyboard. The results revealed a much broader connectivity of various brain areas in handwriting compared to typing. In particular, they noted an increased activity in central and parietal brain regions in handwriting relative to typing, which they link to greater sensorimotor integration. This in turn is argued to be related to the finer motor control required for handwriting compared to typing. The authors relate this to broader educational issues noting that "brain research shows that it is not just any motor activity that facilitates learning, but that accurately coordinating the complex hand movements while carefully shaping each letter when using a pen, is crucial" (p. 2).

One obvious limitation of the study is that the participants were doing single-finger typing with their index finger only. The authors justified this design as desirable "to prevent undesired crossover effects between the two hemispheres" (p. 7), but it does raise the question of whether different, and significant results would have obtained if they had instead measured brain activity in fluent touch typists.

The involvement of multiple areas of the brain instigated by the repetitive hand movements involved in writing has been implicated in the invention of writing itself. Thus Overmann (2016)—and see also Overmann (2022)—has argued (p. 11):

> [W]riting is repeatedly moving the hand to produce marks and visually judging them for legibility in a material that infuences how movements are made and characters formed. Over time, this interaction improves hand–eye coordination, trains the fusiform gyrus to recognize written objects by their features, and increases coordination between the fusiform gyrus and the brain regions that comprehend and produce language and control handwriting movements.

While the invention of writing clearly also involved training the connection between symbol and sound—see Sproat (2023, Chapter 5) for discussion of the neural underpinnings of this connection, Overmann is surely correct that motor movements, facilitating connections between multiple areas of the brain were an important factor.

If Mesopotamian accountants had, instead of styluses, used seals embossed with individual glyphs—a fourth millennium BCE equivalent of movable type—would full writing ever have developed?

3.8 Electronic text and Unicode

The development of digital computers and electronic text encoding systems ultimately led to the redemocratization of script input.

That result was of course not immediate. The earliest computers used limited encoding systems like 7-bit ASCII or EBCDIC, which were designed with limited character sets that were essentially limited to inputting English, and a few other languages that used undiacritized Latin letters. Not surprisingly, as the use of computers spread, the need to develop encoding systems, and associated fonts, that could support other scripts and writing systems, was soon felt. This led to the development of 8-bit encoding systems such as the ISO-8859-X series for various European and Western Asian writing systems; and 16-bit systems—Big5, EUC, GB, JIS and so forth—for East Asian writing systems.[30]

The desire to unify all these systems into a single universal system led to the Unicode Standard, originally developed in the 1980's and promulgated starting in the early 1990's. Unicode's *Basic Multilingual Plane* (BMP) consisting of 65,536 code points covers all modern scripts. However Unicode allows for multiple planes, and ancient scripts such as the cuneiform scripts, or Linear B, have Unicode representations outside the BMP. The Unicode Standard is still under development as further scripts as well as non-linguistic symbol sets continue to be added.[31]

The main advantage of computerization from the point of view of text input, is that it greatly simplifies the work of the typist. For most scripts, one can now easily produce printer quality output with a standard keyboard. The complications that arose for traditional type, hot metal systems, and typewriters for scripts such as Hangul, or the Brahmic scripts, can now be handled by the input system, or by the font itself: TrueType fonts, for example, include rules for rendering combinations of characters, so that composing a complex *akshara* in, say, Devanagari, is handled by the font.

Alternatively, in the case of Hangul, where each syllable combination corresponds to a separate Unicode code point, the input system can handle the combination dynamically. For example if one is using the so-called two-set input system, and one wants to type the two-syllable expression 안해 ⟨an hae⟩ 'don't (do it)', one types the five *jamo* that make up the form in sequence. First ㅇ and ㅏ , which the input system combines into 아. Typing ㄴ converts this into 안. Next one types ㅎ , which at this point the input system thinks is part of the same syllable, so we get 않. But we can forge ahead with ㅐ : since this cannot start a syllable by itself, the input system realizes that the ㅎ must be part of the second, rather than the first syllable, and so this gets corrected to 안해. At each stage, assuming the typeface designer did their job, the output will appear as well-proportioned Hangul syllables.

Even for Chinese, there exist input systems where one can specify a character directly by typing a sequence of keys on the keyboard. One such system

---

[30] See Clews (1988) for an excellent introduction to character set standards.

[31] We note in passing that, unlike older systems like Big5 and GB, Unicode is not itself an encoding system. In GB, the Chinese character for 'one', 一, has the encoding `0xA440`, which means that a text containing that character will have two bytes with the hexadecimal values of `0xA4` and `0x40`. The Unicode code point for 一 is `U+4e00`, but how this is represented in text depends on the particular *encoding* chosen. For example, in the popular encoding system UTF-8, the character would appear as the three-byte sequence `0xe4 0xb8 0x80`.

is Cangjie, where characters are decomposed structurally into a set of simpler components, each of which is indicated by a keystroke. For example the character 法 *fǎ* 'method', is broken down in Cangjie into 水 'water',[32] 土 'earth', and 戈 'spear', assigned to the letters 'e', 'g' and 'i' respectively. Even with this level of specification, however, there are still occasional collisions, where more than one possible character matches the input, which need to be disambiguated by the user.

In any case, far more popular for both Chinese and Japanese are phonetic input methods, typically using pinyin in China, Mandarin phonetic notation in Taiwan, or hiragana for Japanese. Coupled with good language models, such input systems can be highly efficient, but even with such language models they cannot be 100% accurate: they cannot always pick the character that is intended by the writer in every case. This in turn takes us away from script complexity per se, and into the realm of writing system complexity, which is the topic of the next chapter.

With computer input systems we have in effect come full circle when it comes to achieving high quality script forms. With a pen or stylus, the form of the glyphs and their combination was constrained only by the particular writing implement and writing surface used, as well as the skill of the scribe. It was certainly not constrained by dimensionality. With mechanized printing, and with typing, the arrangement of symbols in the script and their manner of combination suddenly became crucial, and depending on the complexity of the script, could be make-or-break for the application of the technology. And even if the technology could be applied, the resulting text was often less than ideal in appearance. But with computer fonts, the work of appearance has been shifted to the typeface designer, and the main skill that the text creator needs is the ability to type. Indeed, with handwriting recognition and speech recognition, one need not even be able to type, merely write in one's script or be able to speak the target language. And with Generative AI, one can if one wishes even dispense with most of the text creation. To varying degrees, modern mechanization has taken on the roles of writing implement, writing surface, scribe and author.

3.9 Script complexity: Summary of the issues

There are many ways to think about the complexity of a script. The simplest is simply to count the number of symbols. This in turn usually correlates with the mean complexity of the characters in terms of the number of strokes that one needs to make in order to produce the glyphs: systems with more glyphs tend to have glyphs that on average are more complex, in order to be able to make the larger number of distinctions apparent to the reader.

---

[32] The three dots on the left of the character 法 are indeed the 'water' semantic radical, a reduced form of 水.

But mean stroke count is most relevant for the legacy technology used to write the script—brushes, pens and styluses, and is only indirectly relevant for later input technologies such as various technologies for printing, and typing. And it becomes even less relevant—for the writer—with computer-based input systems where the correlation between the character's stroke count, and the key presses needed to produce the character is nonexistent.

Script complexity thus must be evaluated as it pertains to the particular technology under consideration. But it also must take another factor into consideration, namely how willing users of the technology are to compromise what would be ideal for the sake of what is practical. Again, if Korea had been willing to fully linearize Hangul, so that 한글 ⟨hangeul⟩ were to appear instead as ㅎ ㅏ ㄴ ㄱ ㅡ ㄹ, then typing in Korean would have been every bit as straightforward as typing in English. But, as we saw, while proposals to linearize Hangul were considered, they were deemed too much of a compromise. On the other hand, in the case of some Brahmic scripts, such as Malayalam, simplification of the script to accommodate the limitations of technology was indeed undertaken.

The compromises needed to accommodate traditional printing and typewriting technologies to non-linear scripts were not inconsiderable. The resulting output was often considered subpar, as we have seen. Thus a fuller characterization of script complexity would be as follows:

> Given ① a particular output technology, and assuming ② a given level of acceptable compromise in the quality of the output, ③ how easily can the script be accommodated by the technology and ④ how easy is it to learn the gestures (Chapter 5) necessary to use the technology?

Take the case of printing in Devanagari, and let us pick Linotype as the technology ①. For ②, we are willing to accept that uncommon ligatures will not be available, and that glyphs that should overlap horizontally will not be properly kerned. For ③, a reasonable question would be whether one can accommodate the needed glyphs on a 90-key keyboard, assuming just one magazine, since switching magazines all the time is costly. Then for ④ the Linotype operator's job is in principle as straightforward as it would be for English, with the proviso that they need to be aware that they are typing 'sectional characters', and how these join up to make a line of type.

Or consider Hangul, and the case of typing ①. If we are willing to accept ② a 'clothesline font' appearance, then ③ one can fit Hangul onto a standard typewriter keyboard, assuming the typewriter has a mechanism to position non syllable-initial glyphs appropriately. The main complexity with the typing gestures ④ is that the typist needs to remember that initials and finals are separate glyphs: the two ㄹ on the keyboard have two different functions, and these functions must be kept straight.

Finally, consider Chinese, and the Double Pigeon 'typewriter' illustrated in Figure 3.21 ①. Here there is in principle no compromise in the quality of the output since this depends upon the quality of the individual type used in the machine; the main compromise ② is that even with a repertoire of

7,000 characters, there may be the occasional rare character that cannot be typed. The board could only accommodate about 2,500 characters, which is not enough to cover everything needed for most genres of text, so without switching boards, or swapping out type, the technology cannot quite accommodate ③ the script. But surely the hardest part is the skill required of the operator ④, which greatly exceeds the skill needed to type in Hangul, much less English.

While it is difficult to provide a rigorous formalization of the above ideas, we propose the following sketch as a basis for such a formalization. In particular:

- ② should be an average 2D edit distance between the ideal presentation and what can actually be achieved—though it cannot be as simple as a pixel-level edit distance.
- ③ could be as simple as what proportion of the glyphs or glyph combinations *cannot* be accommodated in any form on the device.
- ④ is a measure of gesture complexity, though it could be crudely evaluated in terms of the number of positions one must choose from when writing a glyph.

This suggests a formula such as the following:

$$C_{s,t} = \alpha_1 \delta + \alpha_2 \left(1 - \frac{|\sigma|}{|\Sigma|}\right) + \alpha_3 \gamma$$

where $C_{s,t}$ is the complexity for a given script s and technology t; δ is the edit distance; Σ and σ are, respectively, the full set of glyphs and the set of glyphs that can actually be accommodated; γ is the gesture complexity as estimated above; and α_k are normalization coefficients, which are assumed to be values larger than 0, up to and including 1.

For typing English on a traditional typewriter δ would be close to 0 (modulo issues with fixed-width versus variable-width fonts). The second term would evaluate to 0 too since all letters can be accommodated on a standard typewriter keyboard. The number of keys one needs to choose from is 52 (for upper- and lower-case letters only), but that needs to be normalized somehow so as not to dominate the other terms (in machinery/scripts where the other terms are not 0), so arbitrarily set α_3 to 0.01. So then $C_{\text{English,typewriter}}$ would evaluate to $0 + 0 + 0.01 * 52 = 0.52$.

Contrast this with the case of the traditional Malayalam script. Here γ would be similar to that for English since the number of keys on the keyboard is in any case roughly the same. In contrast the second term $1 - \frac{|\sigma|}{|\Sigma|}$ would evaluate to greater than zero since not all of the required glyphs and glyph combinations can be accommodated. As we discussed in Section 3.7.2, the number of consonant conjunct glyphs numbered on the order of about 200 to which one must add the roughly 60 basic letters. Assuming that most vowel diacritics could be handled with dead keys, this would still leave something like 70% of the glyphs or glyph conjuncts that could not be covered by the typewriter, so that the second term would evaluate to approximately 0.7. As for δ, that would be also higher than 0 since the look of the typewriter text is significantly degraded. Supposing one assigned a value of 0.5 to that, and assuming that α_1 and α_2 are both 1, then this would yield $0.5 + 0.7 + 0.52 = 1.72$. The numbers obviously cannot be taken too seriously, but it is in any case clear that a formulation such as the above captures the difference in complexity between the English script, which fits readily on a traditional typewriter, and the Malayalam script, which does not.

Chapter 4
Writing system complexity

4.1 Prelude

In the last chapter we examined how the complexity of a script used to write
a language interacts with technology and thus how it affects the act of writing
using that technology. In this chapter we turn to the complexity of the writing
system. As we introduced earlier, a script is a set of symbols that includes
rules about how those symbols are to be combined to create texts in that
script. A writing system is a *pairing* of a script and a language, a set of rules
that tell one how one should encode the language in the script. In the Latin
script used for English, the conventions of the *script* tell you that in order to
write the word spelled ⟨cat⟩, you linearly arrange the letters ⟨c⟩, ⟨a⟩ and ⟨t⟩
from left to right. On the other hand, it is the spelling rules of the English
writing system that tell you that this is how you spell the word *cat*. For the
Hangul script used for Korean, the conventions of the script tell you that
when you combine the letters ⟨ㄱ⟩, ⟨ㅗ⟩, ⟨ㅇ⟩, ⟨ㅑ⟩, ⟨ㅇ⟩, ⟨ㅇ⟩ and ⟨ㅣ⟩ into
a word, they come out as ⟨고양이⟩. The spelling rules of the Korean writing
system tell you that this is how the word *goyangi* 'cat', is spelled.[1]

Just as the rules for combining the elements of a script together can be
more or less complex, so can the spelling rules of the language. This affects
both reading and writing of course, but since our focus here is primarily
on writing, the issue of interest is how the writer goes from a message that
they wish to communicate to a text that is *correctly written according to the
spelling rules of the language*.

Clearly part of the issue relates to the writer's command of the spelling
rules. Japanese is notoriously difficult and, to a lesser extent, Chinese, since
a user of either of those writing systems must master a system that involves
several thousand characters, and decide which of several in principle pos-

---

[1] Even when there is no conventional spelling system for a language, for example in
informal romanization systems for Arabic dialects, the users of the system must adopt
some conventions for how linguistic information is represented in the script.

sible characters are the correct choices for writing a given word. But English is of course also infamous for having a chaotic spelling system, so much so that it stands out among European writing systems as being the most difficult for children to acquire. Thus, for example, Seymour et al. (2003) report on the reading abilities of children at the end of their first and second years of schooling in thirteen different European national languages. Part of their study examined the accuracy in reading from a list of very familiar words. For the languages with the easiest orthographies in terms of letter-sound correspondence—Finnish, Greek, Italian, Spanish, German, Dutch, Icelandic and Swedish—children were already at a level of 6% error or less by the end of their first year. For the more intricate orthographies, Norwegian-speaking children had 8.2% errors, French-speaking children 20.9%, Portuguese-speaking children 26.5%, Danish-speaking children 28.9%, and English-speaking children 66.1%. The study followed the children into the second year for French, Danish and English: at that point French-speaking children were at under 1% error, Danish-speaking children under 8%, but English-speaking children were still lagging significantly at 23.6% (Seymour et al., 2003, Table 5, p. 153). Clearly English orthography brings with it a cost in terms of the additional effort needed to learn it. English writing may use a small character set, but the rules for using the symbols from those set are often quite messy.[2]

But our main interest is how technology interacts with the writing process, and from that point of view, the cases of English and Chinese or Japanese are obviously very different. If one is typing in English on a computer, as long as one knows how to type, and knows the spelling rules one can—*even in the absence of writers' aids such as spelling correction or word completion*—quite rapidly produce correct text. The English character set is small enough that the decision on how to write a given word can be relegated to the typist—assuming, again, that they know how to spell.

But this is not the case with Chinese or Japanese, at least not on a keyboard with a small number of keys. As we noted in the introduction to Chapter 3, for Chinese there are structural input systems, such as Cangjie, that allow one to input characters based on their structural properties. Other such methods include *Wubi* (Adler, 2020) and the similarly named but distinct *Wubihua* method. With such methods, one can specify a given character directly

---

[2] One curious side-effect of the complexity of English spelling is the uniquely Anglophone institution of the *spelling bee*, a competition that would make no sense for a writing system such as that of Spanish, much less Finnish. In some French-speaking countries *La dictée* takes the place of the spelling bee, but this tests the ability to spell correctly in a dictation context, and thus tests contextual spelling ability. For many years there was also an equivalent test in Dutch `https://en.wikipedia.org/wiki/Grand_Dictation_of_the_Dutch_Language`.

In Japan there is the 日本漢字能力検定 (*Nihon Kanji Nōryoku Kentei* 'Japan Kanji Aptitude Test'), for which the highest level 1 has a pass-rate among native speakers of around 10%. (Source: `https://en.wikipedia.org/wiki/Kanji_Kentei`, retrieved December 28, 2024.)

with a few keystrokes. The downside is that one must learn the required key combinations, which may make these systems less attractive for someone who already knows how to type in an alphabetic script on a standard keyboard. As a result, phonetic input methods—typically *pinyin* in Mainland China, or *zhuyin fuhao* in Taiwan—are more popular. And for Japanese, which already has a native phonetic script in the form of Kana, input systems based on Hiragana are quite standard.

But of course phonetic input systems for Chinese and Japanese have a downside: the mapping between the easily typed phonetic representation of the word, and what one wants to appear on the screen in standard orthography is *not* in general predictable. If one wants the word 'cat' in Chinese one might type its pronunciation in pinyin—*mao*, ignoring tone, which most phonetic input systems for Chinese do ignore—and get 猫 (in Traditional Chinese characters 貓), but the problem is that there are a good many words that have the same pronunciation. Another such word is 毛 'hair' (different tone, but again the input system ignores tone). The same situation obtains in Japanese. One can type *neko* in Romaji, or ね こ in Hiragana, and get 猫. In this particular case one is more lucky: there really are no other Kanji spellings that this could correspond to, though since names of animals are often written in Katakana, one might prefer that the system produce ネ コ, rather than 猫. But in Japanese it is more usual that one must select from a variety of different possible spellings for a given input. In any case, when using a phonetic input system, the decision on how to spell a given word cannot be relegated to the typist: the input system must be involved at least at the level of offering sensible choices to the user. So here the complexity of the Chinese and Japanese writing systems has a more immediate technological impact than is the case in English: in English you can type on a "dumb" terminal, in Chinese or Japanese the system cannot be dumb.

Now the problem with large character sets would appear to apply also to Korean: if one wants to type the word for 'cat' one cannot, on a normal keyboard, type the three syllable-combined glyphs in 고양이. But one *can* type the individual *jamo*, and their combination into syllables is completely predictable. So the issue in the case of Korean is simply one of *script complexity*, not *writing-system complexity*.

Writing-system complexity and its technological implications relates directly to a couple of areas of study, one in the psycholinguistics of reading, and the other in the study of writing systems. We summarize these briefly in the next section.

4.2 Deep and shallow orthographies and logography

One way to think of writing system complexity is in terms of, (a) how much lexical idiosyncrasy there is in determining a word's spelling; and (b) the

amount of context one needs to consider in order to know how to read, or
to spell a word. Consider a writing system like that of Finnish, where there
is an almost one-to-one mapping between the phonemes that make up the
pronunciation of a word, and the letters used to spell it. In such a case, if
one wants to spell a given word, then one merely needs to consider its pronunciation, and spell accordingly. Alternatively, reading a word on a page
merely requires the relatively straightforward application of a simple set of
letter-to-sound rules. One need not know anything about the word *Suoma-*
lainen 'Finn' other than its pronunciation (/ˈsuɔmɑlɑinɛn/), to know that it is
spelled ⟨Suomalainen⟩.

Obviously such a scenario is not the case for English. Not only is there
much lexical idiosyncracy—the word *debt* is spelled with a 'silent' ⟨b⟩, for
reasons that will only appeal to an etymologist—but one often needs to consider the broader context when determining whether the word pronounced
/ðɛə/ should be spelled ⟨there⟩, ⟨their⟩ or ⟨they're⟩.

This ties in with a well-studied theme in the psycholinguistics of reading
that we have already seen, namely the distinction that we touched on in
Section 2.3.2 between *shallow* versus *deep* orthographies. Again, in simple
terms, shallow orthographies comprise writing systems where there is a fairly
direct relationship between how a word is pronounced and how it is spelled.
Deep orthographies, in contrast, are writing systems where the relationship
is much more indirect. The focus of much of this research has been on how
the two types of orthography affect the reading process, as well as how they
affect learning to read. See Besner & Smith (1992); Katz & Frost (1992); Ellis
et al. (2004); Miller et al. (2014) for psycholinguistic work, and Bosch et al.
(1994); Marjou (2021) for computational interpretations.

To some extent, the deep-shallow distinction implicitly makes a rather bi-
ased assumption: the simplest, shallow, orthographies are taken to be those
where one can pronounce a written word simply by applying some simple
letter-to-sound rules. In other words it is taken as a given that the ideal orthography is one that is as close as possible to being essentially a phonetic
transcription of the pronunciation of words. However, while all fully functional writing systems must encode sound, this does not mean they must
only encode sound, or that they need do it in a completely regular way. From
the writer's point of view, this may indeed be ideal, but from the reader's
point of view, additional cues that might help identify the particular word
intended can often be useful. Japanese writing offers a clear illustration of
this point. In Japanese, while many words have Kanji spellings (sometimes
more than one), all words can in principle be written in one of the two Kana
systems. Indeed it is not at all unusual to find within the same text, a word
written both in its canonical Kanji spelling, and in Hiragana or Katakana.
One can write whole sentences in Hiragana, and one often finds such sentences on signs that are intended to be readable by children. For example,
the Tokyo Metro displays the message in Figure 4.1 on the doors of its trains.
Apart from the word for 'door' *doa*, which is spelled in Katakana, the entire

Fig. 4.1 Warning message on the doors of a Tokyo Metro subway car. The text, ひらくドアにこちゅういください, *hiraku doa-ni gochūi kudasai*, means 'beware of the opening doors'. Source: own work.

sentence is written in Hiragana, again because it is intended to be readable by children. From the writer's perspective, this way of writing the sentence is very straightforward, since it is a fairly simple matter to spell in Hiragana given the pronunciation of the words. But from the mature reader's perspective, this written representation is not necessarily the easiest, especially given the lack of any spacing to separate words in Japanese. For reading, the more standard spelling whereby the words for 'open', and 'beware' are written partially in Kanji is arguably easier:

開くドアにご注意ください

The dual issues of the amount to which one finds idiosyncratic spellings, and the amount of context one needs to determine how to read or spell a word, relates also to the notion of *logography*, introduced in Chapter 2. However, here we need to be a bit clearer about what the term 'logography' means. In the earlier discussion we introduced the term as denoting systems where a symbol stands for a word or morpheme, without reference to the word's pronunciation. Think again of the Chinese character 馬 *mǎ* 'horse', which represents the whole morpheme/word for 'horse', and cannot be broken down into parts that represent the pronunciation.

But this can in turn be related to the two criteria introduced above: first of all 馬 is an idiosyncratic spelling: one simply has to know that this is how the word is written. And it is also contextually dependent, at least from the point of view of determining how to spell, since 馬 is not the only way to write words that have the pronunciation *mǎ*.

Scholars of writing systems tend to think of cases like 馬 when they think of logography: a single symbol that represents a word or morpheme. But this is actually a rather parochial view, since there are many cases where it makes sense to think of a string of *multiple* symbols as logographic. Consider the case of the monomorphemic Japanese word 百足 *mukade* 'centipede'. This is written with two Kanji that happen to mean 'hundred feet', but this spelling cannot be broken down in any way that corresponds to the pronunciation of the word: One cannot break it down into, say, 百 *mu* 足 *kade*. In this case the characters in combination are being used purely for their meaning to spell the word for 'centipede'. This would seem to fit the definition of 'logography' insofar as the two characters together represent a word/morpheme without reference to sound. But that this is not the case of a *single* symbol representing a word, but rather two symbols. A similar case involving three Kanji is 秋刀魚 *sanma* 'saury', where the Kanji together mean 'autumn sword fish', picked presumably because saury are typically fished starting in the autumn, and are long slender silver fish resembling the blade of a sword. Again, this spelling is unrelated to the pronunciation of the word.

Now to be sure, while 百足 or 秋刀魚 can be said to be logographic spellings for *mukade* and *sanma*, respectively, it is also true that the individual glyphs in these cases are also logographs. Thus, apart from their use to write 'centipede' 百 'hundred' and 足 'foot', are themselves logographs representing the words for 'hundred' (typically *hyaku* in Japanese) and 'foot' (native reading: *ashi*).

But do the basic symbols themselves need to be logographs, in order to consider a multi-symbol spelling to be logographic? If it were decided that henceforth the English word *dog* were to be spelled *zgi*, a spelling which obviously bears no relation to the pronunciation of the word, could one consider this to be a case of logography? In fact a case quite similar to this obtained in the writing systems of Middle Persian languages. Aramaic was the administrative language of the Persian empire, and as a result of this, the Aramaic script was also adapted as the basis of the writing system of Middle Persian languages. Since the scribes were also literate in Aramaic, they adopted the practice of writing Persian words with the Aramaic spellings of words with the same meaning. Thus the Persian word *šax* 'king' could be spelled as ⟨mlkʔ⟩ (Skjaervo, 1996), reflecting the Semitic etymon for 'king'. But despite the Aramaic spellings, it was clear from the syntactic and morphological context that such words were to be *read* as Persian words. Given that Aramaic was a phonemic writing system, where the scribes could have adopted a spelling such as ⟨šax⟩ for 'king', this may seem like an odd choice. But from the scribes' point of view this made sense: they already knew how to spell a word with the meaning 'king', so rather than adopt a spelling based on the Persian reading, why not instead use the spelling they were familiar with. In fact, the situation was no different from the case where Japanese scribes, using the Chinese script to write their own language, started adapting characters to write Japanese words with the same meanings (Handel, 2019). The character 王 (Mandarin *wáng*) meaning 'king', could also be used to represent the native Japanese word for 'king' (Old Japanese *ofokimi*).

While some scholars of writing systems may disagree with the characterization of some of the cases above as constituting logography, it must be stressed that the literature on writing systems has not exactly provided a clear definition of what the term means. Sproat & Gutkin (2021) surveyed seventeen definitions of the term, starting with that of Gelb (1952) ("Lo-

gograms, that is signs for words of the language.", p. 99), through that of
Handel (2019), pages 7–8:

> In a logographic system, the basic graphic elements represent meaningful ele-
> ments of the spoken language, so that identically pronounced but semantically
> contrastive elements have distinct graphic representations.

Sproat & Gutkin (2021) note that Handel's description of the meaning of the
term is probably also the closest in the literature to a formal definition. And
it also relates nicely to the second aspect of logography introduced above,
namely how much context one needs to consider in order to know how to
spell a word.

As the second author argued in previous work—Sproat (2000), and see
also Rogers (2005)—logography is not an all-or-nothing proposition: a writing system is not either logographic, or phonographic. Rather logography is
a matter of degree, and relates directly to the proportion of idiosyncratic
spellings, and the degree to which one has to consider the context in determining how to spell a word. This implies that one could measure the degree to
which a writing system is logographic. Such a measure should place systems
like those of Chinese or Japanese fairly high on the scale; but systems like
Finnish, which has a very regular and simple segmental phonographic writing
system, low on the scale. Sproat & Gutkin (2021) argue that a reasonable
measure of the degree of logography is the amount of context around a word
that is required to know how to spell it, given the pronunciation. If you are
told to spell a word pronounced /ɹiːdz/, you cannot be sure if the word intended
is to be spelled ⟨reads⟩ or ⟨reeds⟩. But if given the context *along the lake the
____ were undulating in the wind*, it is clear which spelling is the intended
one. Sproat & Gutkin (2021) propose as a measure the proportion of the
attention that a neural sequence-to-sequence model pays to material outside
the target word. In an ideal phonographic system like that of Finnish, the
model would only need to attend to the word itself in order to know how to
spell it. But for English, and even more so for Chinese or Japanese, the model
would often need to attend to material outside the word. As Sproat & Gutkin
(2021) show, this rather natural computational interpretation affords a measure that corresponds very closely to intuition about how logographic—and
how shallow or deep—a writing system is. And this in turn corresponds to
the notion of writing-system complexity: a complex writing system requires
one to consider more of the context than a simple writing system.

How does all of this relate to text input? As we noted earlier, if one is trying
to input text in a language with a complex writing system and the size of the
writing system's alphabet is small enough that it can fit on the keyboard,
then the complexity of the writing system is in principle a problem only for
the typist. Here, the typist must simply be aware of the spelling system of
their language, and attend to make sure that they are not spelling ⟨there⟩
when they meant ⟨their⟩. If on the other hand, as is the case in Chinese or
Japanese, the character set is too large to fit on a standard keyboard and one

is therefore using a phonetic-based input system, then the problem becomes a joint task between the machine and the writer. The machine—a computer or hand-held device—must have a good enough model of the written language in order to be able to offer plausible spellings given the phonetic input. And the writer must constantly attend to what the machine is doing to make sure that mistakes are not introduced. In any case, even for easily typable writing systems, there are computational aids which have been growing in sophistication over the last half century. To these we now turn.

4.3 Tools for complex writing systems

For most of history, the technology used to input text was passive. Styluses, brushes, pens, woodblocks, movable type and typewriters were tools used by writers to transfer text to a writing surface, but what ended up on the surface was entirely within the control of the writer.

With the advent of computer-based input systems, this all changed. Now, when one writes one can get aids ranging from suggestions from the system for alternative spellings; to autocorrect features that will make these corrections without asking the user; to, in the limit, generative AI systems that can compose the text entirely for the writer, so that the notion of 'writer' itself becomes unclear.

In this section we briefly review some of these input systems, starting from the second half of the 20th century, whose purpose is to aid the writer in dealing with the complexities of the writing system they are tasked with composing text in.

4.3.1 Early writer's tools

One of the earliest tools to help writers with a complex writing system— English—was the UNIX Writer's Workbench (Gehani, 1987), developed at AT&T Bell Laboratories by Lorinda Cherry and Nina McDonald in the early 1980's. The UNIX Writer's Workbench was a suite of tools including various grammar checking components, a spell-checker, and even a component called 'sexist', which attempted to find potentially sexist terminology in documents. The SPELL component was originally developed by S. C. Johnson and further developed by Doug McIlroy (McIlroy, 1982).[3] It included a dictionary, and a rather ad-hoc collection of morphological rules that was designed to cover words traditionally not included in dictionaries—e.g., regularly in-

---

[3] The GNU system's SPELL is an open-source version of the original Unix SPELL system. An earlier spell checker written in assembly language for the DEC PDP-10 is credited to Ralph Gorin at Stanford in 1971 (Peterson, 1980).

flected forms like *carrots*—as well as to reduce the size of the dictionary as much as possible, space and search time being a major concern for computer hardware of the early 1980's. Sproat (1992), Chapter 1, discusses SPELL's morphological decomposition approach in more detail.

The Writer's Workbench tools were offline systems in that one submitted a completed document to the tools, which would flag the text for possible problems. The writer would then evaluate the output of the system and make corrections as needed. Online systems, that flagged possible misspellings while one was preparing the text came later, e.g. with Microsoft Word 95.

4.3.2 Early Chinese and Japanese input systems

Phonetic input systems for Chinese (using the Mandarin phonetic system *zhuyin fuhao* in Taiwan, or the Roman-script-based *hanyu pinyin* in China), or Japanese (Hiragana or Romaji) were first developed in late 1970's and early 1980's.

The simplest approach is to have the user type the pronunciation of a desired character, and then present a list of characters that have that reading. If the characters are presented in descending order of frequency, the desired character will be among the first few of those presented most of the time. In practice, good typists could input text very quickly using such systems.

However, the system could obviously be improved by including a comprehensive dictionary. While in general there will be many characters that correspond to a particular phonetic transcription, there are far fewer words. The Chinese syllable *ma* (not including tone) could be any of 馬, 麻, 碼, 媽, 螞, among many others; whereas, the word *mayi* is most likely to be 螞蟻 *mǎyǐ* 'ant', with few other reasonable competitors. There is however one issue with word-based input. Unlike many modern writing systems, neither the Japanese nor Chinese writing systems use space to delimit words. As a result, users may be unsure what constitutes a word, and in general their conception of wordhood may differ from that of the creators of the dictionary.

One way around this is to dispense with a lexicon entirely and propose candidates based on a statistical *language model* that, given an input pronunciation, proposes a rank-ordered list of character proposals, given the input and the preceding context. Thus in Mandarin, a syllable *feng* given the previous context 臺灣有颱— *táiwān yǒu tái—*, is most likely to correspond to 風 *fēng*—thus 臺灣有颱風 'Taiwan has typhoons'.[4] This approach obviates the need for the writer to know what the input system designer considered to be a word: instead they can simply write 'character-by-character' with a reasonable chance of the system selecting the desired target character most of the time.

---

[4] The second author designed an early version of such a system (Sproat, 1990).

4.3.3 More on spelling correction

As computers became more powerful, and online resources such as dictionaries and text corpora became more widely available, it was possible to develop systems that went well beyond, and were easier to develop, than early hand-engineered systems such as the SPELL system introduced above. Indeed, spelling correction was for many years a popular niche area of Natural Language Processing research.

A recent review of spelling correction systems from the early 1990's through 2019 can be found in Hládek et al. (2020). Spelling errors are traditionally classified into two main categories, namely typographical errors, and cognitive errors (Kukich, 1992; Toutanova & Moore, 2002; Pirinen & Lindén, 2014). Typographical errors are, of course, due to errors in execution of text input on a keyboard. These are influenced by factors such as skill of the operator, distraction and time pressure, but do not directly relate to the complexity of the writing system. Typographical errors typically create words that are not words of the language, such as ⟨teh⟩ (for ⟨the⟩), though by accident they may produce words that are correctly spelled, but not the intended word.

Cognitive errors, on the other hand, relate directly to the complexity of the writing system. They may involve picking an inappropriate homophone—⟨there⟩ instead of ⟨their⟩, or if the writer is a bad speller, may produce tokens that are not words at all, such as ⟨stumik⟩ for ⟨stomach⟩. The chosen spelling will often be phonetically reasonable in that the intended word *might* have been spelled that way or approximately that way.

Before proceeding further, it is worth thinking a bit about the logical structure of the spelling-correction problem. This is often characterized as a *source-channel* problem. The text one is trying to recover—the *source*— is assumed to be correctly spelled, but it has been corrupted by passing it through a noisy *channel* to produce an actual text that has misspellings, typographical errors and so forth. The problem then is of understanding the workings of the noisy channel. One of the components is typographical errors, where errors in keystrokes produce garbled output; and another is cognitive errors, where the error is not due to mistyping, but rather to a misunderstanding or confusion on the part of the writer. In either case, in reconstructing the corrected source, for any given misspelled/mistyped word, one needs to consider two questions:

- What is a reasonable set of candidate correctly spelled words?
- And, which of these is the most appropriate given the context?

Cognitive errors in spelling are, needless to say, more difficult for an automated spelling-correction to solve than typical typographical errors that involve transpositions, insertions or deletion of letters: It is usually easy to detect a non-word such as ⟨tehre⟩ and propose corrections from a set of typographically close spelled words.

Similarly, if the cognitive error involves writing a non-word such as ⟨stumik⟩, then it is relatively easy for a system to detect that this probably should be corrected. However, unlike the case of typical typos, the correct spelling is not necessarily going to be close typographically. Solutions to this problem generally have involved finding words that are close phonetically. ⟨stumik⟩ is a possible spelling of /ˈstʌm.ək/, which would normally be spelled ⟨stomach⟩. Techniques for finding words that are spelled differently but are close in phonetic space actually have a long history. The first such method was Soundex (Russell, 1918, 1922), which was originally developed as an indexing system for grouping together names that are spelled differently, but are similar in pronunciation. In the original Soundex system, names were indexed by their first letter, and the remaining letters were mapped to one of eight roughly phonetic categories, removing duplicates from the same category. The name *Hopp* was mapped to *H12*: all vowels and ⟨y⟩ are mapped to '1', and ⟨p⟩ is mapped to '2', the second '2' being removed since it is a duplicate. *Haup* was mapped to the same *H12* sequence, in this case with the duplicate '1' (for ⟨u⟩) being removed. ⟨F⟩ is also in the same category as ⟨p⟩, so *Hopf* would also be mapped to *H12*. Soundex was a good way of grouping names so that roughly similar names would end up in the same part of a card filing system such as those that used to be used in libraries, but it has a number of drawbacks for serious use as a way of computing phonetic similarity. Words that are grouped together are not necessarily that similar in pronunciation: *Heap* would also be coded as *H12* though it is rather unlikely that someone would misspell *Hopp* as *Heap*. Conversely, words that *are* phonetically similar, or identical, but which begin with different letters, will have different codes; *Knot* and *Nott* are identical in pronunciation, but the former is *K714* whereas the latter is *N14*. Nonetheless, as Hládek et al. (2020) discuss, a number of spelling correction systems have used Soundex, or variants thereof, as a phonetic similarity metric, possibly in combination with other similarity metrics.[5]

In cases where the writer produces a non-word, the task of the system is to derive a set of possible corresponding correctly spelled words, and then either correct the spelling to the most likely candidate given the context, or suggest to the user possible corrections with the most likely candidates ranked at the top of the list. This requires that the spelling correction system have a good model of the language so that contextually appropriate candidates are ranked higher than ones that are unlikely given the context. For this purpose, spelling-correction systems have used various statistical language modeling approaches, akin to those that we discuss elsewhere in the book.

A more difficult case is where the writer produces a misspelling that is actually a valid word, but one that happens to be inappropriate to the context. Often these sorts of misspellings will involve words that are phonetically

---

[5] One advantage of Soundex over more sophisticated methods such as full *grapheme-to-phoneme conversion* system is that it is exceedingly efficient to compute a Soundex code for any given word.

identical or similar, so the same phonetic models come into play. But there
is an additional problem that the system needs to notice that the word is
out of place, and only then proceed to suggest alternatives. Clearly this has
to be done with some care, since one does not want a system that willy-
nilly corrects errors that are in fact not errors. To give an example, if I type
They're dog has spots, ⟨they're⟩ is clearly not appropriate given the context.
In this case there are only two words that are phonetically similar enough to
count as candidates, and only one of these (⟨their⟩) that makes sense. False
positives are a serious issue especially when it comes to proper names. Those
unfamiliar with the Church of Jesus Christ of Latter-Day Saints' toponym
Deseret, might feel inclined to view the newspaper title *The Deseret News*
as a typo for *The Desert News*. Detecting and correcting—but not *miscor-
recting*—this kind of error is particularly tricky. An early attempt to address
this problem robustly was Golding & Roth (1998)'s Winnow-based spelling
correction system, which was able to detect and propose corrections for such
errors with about 96% accuracy for a list of about 20 'confusion sets', i.e.
words that could be confused for one another in writing.[6]

However, it was not really until the introduction of Deep Learning that
fully automated spelling correction that captured all kinds of errors became
safe, or at least safe enough. As Hládek et al. (2020) document, a variety of
machine learning and language modeling approaches have been applied to the
problem of spelling correction over the last three decades, but "progress of
spelling correction ... was slow until the introduction of deep neural networks"
(p. 20).

Still it is fair to ask: with the advent of Large Language Models (LLMs),
has the problem of writing system complexity as a challenge for the writer
been effectively solved? How well do such systems work?

Consider first a case where, due to cognitive errors, the system needs to
correct seriously misspelled words. How deviant can the spelling be, and still
have the system offer a corrected rendition? As a simple test consider the text
from the 19th century political cartoon presented in Figure 4.2 lampooning
racist Jim Crow voter restriction laws:

> Eddikazhun Qualifukazhun. The Blak man orter be eddikated afore he kin vote
> with us Wites, signed Mr. Solid South.

As of late 2024, before the rollout of Gemini,[7] the spelling correction system
underlying Google Docs did not do very much with this example. It flagged
only two of the words, namely *Blak*, for which it correctly proposed *Black* as
a correction; and *orter* for which it misproposed *porter*.

---

[6] Not all of Golding and Roth's confusion sets were spelling errors, *sensu stricto*, as they
included pairs like *amount/number* and *I/me*.

[7] `https://gemini.google.com/app`

Fig. 4.2 A 19th century political cartoon criticizing literacy tests for Black voters in the Jim Crow South, depicting an ill-educated White Southerner writing a poster with the text: "Eddikazhun Qualifukazhun. The Blak man orter be eddikated afore he kin vote with us Wites, (signed) Mr. Solid South." Source: `https://en.m.wikipedia.org/wiki/File:The_color_line_still_exists%E2%80%94in_this_case_cph.3b29638.jpg`, originally from *Harper's Weekly*, vol. 23 (1879), Jan. 18, page 52. Image is in the public domain in the United States because it was published (or registered with the U.S. Copyright Office) before January 1, 1929.

On the other hand LLMs seem to be able to handle even this degree of deviance. For example, the online interface of ChatGPT 4o mini (`https://chatgpt.com/`), renders the text perfectly:[8]

> "Education Qualification. The Black man ought to be educated before he can vote with us Whites, signed Mr. Solid South."

---

[8] We used the prompt:

The following text is misspelled: "Eddikazhun Qualifukazhun. The Blak man orter be eddikated afore he kin vote with us Wites, signed Mr. Solid South." Please correct it.

All accesses of the online interfaces to LLMs were conducted in November 2024. The behavior may have changed subsequent to our attempts.

We cannot of course rule out that ChatGPT has seen this particular example, which is famous, in its training data. However a similar test with a different text that we constructed—"You orter no that only eddikated peeplez kin vote in this state"—also produced the correct output: "You ought to know that only educated people can vote in this state."

We obtained a similar result with Gemini, suggesting that in general LLMs are capable of handling even highly deviant spellings.[9]

Consider now a case involving substitutions of contextually inappropriate real words in place of homophonous words:

> Correct any misspellings in the following sentence: "I red a column in The Desert News out of Salt Lake City the other day, and it sed that several deseret species are going extinct."

ChatGPT correctly rendered this as follows, even correctly offering to italicize the title of the newspaper:

> Here is the corrected sentence:

> I read a column in *The Deseret News* out of Salt Lake City the other day, and it said that several desert species are going extinct.

Gemini did not do as well. Oddly it noted that *deseret* (in *deseret species*) was misspelled, it did not actually correct it:

> Certainly! Here's the corrected sentence:

> "I read a column in The Desert News out of Salt Lake City the other day, and it said that several deseret species are going extinct."

> The words "red," "sed," and "deseret" were misspelled.

With ChatGPT we tried a similar case in Japanese for the sentence 次の駅に電車が停まりません。 *tsugi no eki-ni, densha-ga tomarimasen* 'the train does not stop at the next station.' In the prompt we misspell the words for 'station' 駅 *eki*, and 'stop' 停まりません *tomarimasen*. In the first case we substitute the homophone 液 *eki* 'liquid', which does not make sense in the context, and in the second 泊まりません *tomarimasen*, which means 'stop at', but in the sense of staying at a lodging. We underline these in the example below, but note that the LLM was not given any hints about which words were misspelled. The prompt before the example sentence, reads "The following sentence is incorrectly written: please improve it.":

> 次の文は間違って書かれている。改善してください

> 「次の<u>液</u>に電車が<u>泊</u>まりません。」

---

[9] This is fortunate. Given recent right-wing attacks on education, especially in the United States, and the concomitant elevation of ignorance over knowledge, future generations may need all the help they can get even with such matters as basic literacy.

Model	# Parameters	# Errors	Total	Error Rate
Mistral 7B Instruct v0.3	7b	41	354	0.12
Qwen QwQ 32B Preview	32b	31	354	0.09
Llama 3.3 70B Instruct	70b	23	354	0.06
Qwen2.5 72B Instruct	72b	19	354	0.05
GPT-4o-mini	??	17	354	0.05
Claude 3.5 Sonnet	??	6	354	0.02

Table 4.1 Performance of various LLMs on the small set of misspelled English sentences. Number of parameters is reported for all open-source models. See Appendix 4.A for model details and list of example sentences.

ChatGPT corrected this appropriately, and as part of its feedback, explained that "in this fashion, substituting 駅 'station' for 液 'liquid', the meaning becomes clear. This is a sentence that talks about a station where the train does not stop." While it did not explicitly mention the misspelled 泊まりません, it nonetheless corrected it:

「次の駅に電車が停まりません。」

このように「液」を「駅」に修正すると、意味が明確になります。電車が停まらない駅について話している文になります。

Obviously testing a few examples does not demonstrate that LLMs can handle *any* misspelling in any context. To conclude this section, we report on a more systematic test with six language models of varying numbers of parameters for 20 misspelled example sentences each from English and Korean. We summarize the results here, and provide details on models and examples in Appendix 4.A at the end of this chapter.

Starting with English, we compared a range of open-source and proprietary LLMs on the task of correcting massively misspelled English sentences. Models ranged from 7-billion to 72-billion parameters for the open-source models, and are probably in the range of several hundred billion parameters in the case of the two proprietary models (Anthropic's Claude and OpenAI's GPT). More details on the models, the prompt and the specific examples are found in Appendix 4.A, as well as the correct spellings of each example sentence and some sample miscorrections.

For all model outputs, the word edit distance—substitutions, deletions and insertions—were computed and reported as an error rate relative to the total number of words in the hand-corrected set. Results are reported in Table 4.1. As can be seen, there is a fairly straightforward negative correlation between error rate and model size.

Korean examples were processed similarly (see Appendix 4.A for details) and results are presented in Table 4.2. The overall trend of larger models performing better, which we observed for English, also carries over to Korean, though Llama performs unexpectedly worse than the smaller Qwen 32B model. However, the error rates are in general much worse than for English.

Model	# Parameters	# Errors	Total	Error Rate
Mistral 7B Instruct v0.3	7b	78	130	0.60
Qwen QwQ 32B Preview	32b	65	130	0.50
Llama 3.3 70B Instruct	70b	80	130	0.62
Qwen2.5 72B Instruct	72b	59	130	0.45
GPT-4o-mini	??	46	130	0.35
Claude 3.5 Sonnet	??	26	130	0.20

Table 4.2 Performance of various LLMs on the small set of misspelled Korean sentences. Number of parameters is reported for all open-source models. See Appendix 4.A for model details and list of example sentences.

This presumably relates in part to there being generally less data for Korean than for English. While we do not know the amounts of data that the various models have been exposed to—particularly for the large proprietary models, this information is not published—we might use the number of Wikipedia pages as a proxy for the relative amounts of data to which the models might have been exposed, though very likely the true amounts will be much more skewed in favor of English than Wikipedia content would suggest. According to Wikipedia's own statistics,[10] there are 6,989,228 pages for English, and 702,838 for Korean, a ten-fold difference. We also observe that many of the Korean errors involve differences in spacing—whether to consider a given string as one word or two—something that even Korean speakers are often inconsistent about.

The stark performance gap between English and Korean suggests that the problem of unrestricted spelling correction is not solved in any general sense by LLMs—merely, perhaps, for those languages for which truly massive amounts of data are available. A major gap remains between performance on very-high resource languages like English, and medium-resource languages like Korean.

4.4 Summary

Writing itself is a technology—the first speech-and-language technology ever developed (Sproat, 2010). But for writing to exist, other technologies—writing tools—were needed. For most of history, writing tools were passive, and as such could do little to help the writer who made mistakes in the written forms of words. With the advent of computers, this changed, and machines have become increasingly useful tools as their own knowledge of language increased.

In their training data, today's LLMs have seen text in quantities that would take a human thousands of years to read. The knowledge incorporated

---

[10] https://en.wikipedia.org/wiki/List_of_Wikipedias, accessed May 3, 2025.

into these models is so good that helping with the intricacies of the English or Japanese writing systems are among the simpler of the tasks that they can perform. Nonetheless, as the results in the Appendix highlight, writing system complexity remains a problem for systems that would aid any writer of any language in dealing with the intricacies of their standard orthography.

4.A Appendix: LLMs and spelling correction

4.A.1 English examples

The model was prompted with the following instructions:

```
You are a helpful English teacher.

The following 20 sentences may contain misspelled words,
or words that are spelled in an unconventional way.
Please correct the sentences to standard American
English spelling. In your response, please retain
the numbers (1., 2. etc.) with each sentence and
put your entire set of corrected sentences within
an <OUTPUT></OUTPUT> tag:
```

after which it was presented with the following twenty sentences:[11]

1. I cain't stumik it ennymoor.
2. Do yu think that its eezy?
3. I blive Kinnidy wen he sez flooride in watter kin effect chillerns innerlekshul developmunt.
4. Happy to liv in tha USA, wher we consentrate on tha funnermennals, like reading, riting un rithmatic.
5. Wen Eelon Mask meks owr gubmint smaller, all squirls ll feel safer.
6. Thanks to Justis Thomas therty five yerz uhv pubic survice, we kin shoot up a skul eny time we want.
7. Trump likes unedderkated peeple like me.
8. Chocklit iz my faverit flayver.
9. The Orijun ov Speesheez shud be band koz it goze agin the Bibul.
10. Kenidy iz gonna protect owr preshus boddily flooids.
11. There meting hymn inn Gnu Deli.
12. Nah then, Freddy: look wh' y' gowin, deah.
13. Will ye-oo py me f'them?
14. I think and evrything that happins to me but I cant think anymor because I have nothing to rite so I will close for today... yrs truly

---

[11] Sources for a subset of the sentences: 12-13, *Pygmalion*, George Bernard Shaw; 17-18, *David Copperfield*, Charles Dickens; 19, *Feersum Endjinn*, Iain M. Banks; 14-16, 20, *Flowers for Algernon*, Daniel Keyes. Note that we assume that at least some of the examples from literature—though mostly in their original rather than correctly spelled form—will have been seen by various LLMs in their training data hence the need to provide made up examples.

15. Dr Strauss says I shoud rite down what I think and remembir and evrey thing that happins to me from now on.
16. dr Strauss and perfesser Nemur I cant rite good but he says it dont matter he says I shud rite just like I talk and like I rite compushishens in Miss Kinnians class at the beekmin collidge center for retarted adults where I go to
17. Theerfur 'tan't my intentions to moor Missis Gummidge 'long with them, but to find a Beein' fur her wheer she can fisherate for herself.'
18. "Wheerby," said Mr. Peggotty, "my sister might—I doen't say she would, but might—find Missis Gummidge give her a leetle trouble now-and-again."
19. Thi hydravater car is 1 ov thi old wooden 1s wot kreeks a lot & it smelz ov rope-oyl & varnish & thi empty watir tanks underneeth thi deck maik big boomy spooky noyses as it climes up thi wol ov thi hol.
20. Mr Donner gives me 11 dollers a week and bred or cake if I want. I am 32 yeres old and next munth is my brithday.

Corrected versions of the sentences are as follows:

1. I can't stomach it anymore.
2. Do you think that it's easy?
3. I believe Kennedy when he says fluoride in water can affect children's intellectual development.
4. Happy to live in the USA, where we concentrate on the fundamentals, like reading, writing and arithmetic.
5. When Elon Musk makes our government smaller, all squirrels will feel safer.
6. Thanks to Justice Thomas' thirty-five years of public service, we can shoot up a school any time we want.
7. Trump likes uneducated people like me.
8. Chocolate is my favorite flavor.
9. The Origin of Species should be banned because it goes against the Bible.
10. Kennedy is going to protect our precious bodily fluids.
11. They're meeting him in New Delhi.
12. Now then, Freddy: look where you're going, dear.
13. Will you pay me for them?
14. I think and everything that happens to me but I can't think anymore because I have nothing to write so I will close for today... yours truly
15. Dr. Strauss says I should write down what I think and remember and everything that happens to me from now on.
16. Dr. Strauss and Professor Nemur I can't write well but he says it doesn't matter he says I should write just like I talk and like I write compositions in Miss Kinnian's class at the Beekman College Center for Retarded Adults where I go to

17. Therefore 'tain't my intentions to moor Mrs. Gummidge 'long with them, but to find a Being for her where she can fisherate for herself.
18. "Whereby," said Mr. Peggotty, "my sister might—I don't say she would, but might—find Mrs. Gummidge gives her a little trouble now-and-again."
19. The hydrovator car is one of the old wooden ones that creaks a lot and it smells of rope oil and varnish and the empty water tanks underneath the deck make big boomy spooky noises as it climbs up the wall of the hall.
20. Mr. Donner gives me 11 dollars a week and bread or cake if I want. I am 32 years old and next month is my birthday.

The LLMs used were:

- Mistral 7B Instruct: `https://huggingface.co/mistralai/Mistral-7B-Instruct-v0.3`. The Mistral AI Team.
- Qwen QwQ 32B Preview: `https://huggingface.co/Qwen/QwQ-32B-Preview`, (Qwen Team, 2024b).
- Qwen2.5 72B Instruct: `https://huggingface.co/Qwen/Qwen2.5-72B-Instruct-GPTQ-Int4`, (Qwen Team, 2024a).
- Llama 3.3 70B Instruct: `https://huggingface.co/unsloth/Llama-3.3-70B-Instruct-bnb-4bit`, Meta and Llama Team
- Claude 3.5 Sonnet: `us.anthropic.claude-3-5-sonnet-20240620-v1:0`, Anthropic.
- GPT-4o-mini: OpenAI

Some sample miscorrections follow:

- Mistral 7B Instruct: *I believe Kennedy → I believe Cindy*
- Mistral 7B Instruct: *Thanks to Justice Thomas' thirty-five years of public service → Thanks to Justice Thomas's fifty-year ruling*
- Qwen QwQ 32B Preview: *Trump likes uneducated → Trump likes undereducated*
- Qwen QwQ 32B Preview: *They're meeting him → Their meeting hymn*
- Llama 3.3 70B Instruct: *all squirrels will feel safer → all girls will feel safer*
- Llama 3.3 70B Instruct: *where I go to → where I go to learn*
- Qwen2.5 72B Instruct: *Now then, Freddy → Nah then, Freddy*
- Qwen2.5 72B Instruct: *where she can fisherate for herself → where she can fish for herself*
- GPT-4o-mini: *I can't stomach it → I can't stum it*
- GPT-4o-mini: *They're meeting him → Their meeting hymn*
- Claude 3.5 Sonnet: *The hydrovator car → The hydraulic car*

4.A.2 Korean examples

For Korean we used the following twenty examples:[12]

1. 깟뻬뜨랑 리뿔이 뜨럽고 므리커럭이 게쏙 냐와쑈 우뤼갸 청쑈섀로다혀뚜여.
2. 쑥쏘까많이낙후뙤어있꼬엘베없꼬4층이라짐많으면깨꼬쌩합니따.
3. 화짱씰많이 낡았씁니따.
4. 거런데 위불둘이 넘흐 멀휘카락 붙어있고 얼룩이쏘 굴헤요.
5. 건무울도 오뤠대씀.
6. 뙨쏘리를많히쓰꼬모엄을빠꾸는께꾸칙이다.
7. 멘날 하라버지랑 놀고 시따.
8. 희안하게 몇일 지나면 금새 이저요.
9. 감기 낳으세요.
10. 뒤짜석 벨트 꼭 메세요.
11. 요즘 어의엄는 사고가 만테요.
12. 산뿔이 겉잡을 수 업씨 번졌다.
13. 마마잃은 중천공이다.
14. 밥도 안주고 잠도 안 제우는 건 가오캥이 아니냐?
15. 나한테 일해라 절해라 하지마.
16. 이 옷 왜숭모가 사주셔써.
17. 이 돈이 내가 낼 쑤 인는 맥심원이야.
18. 다르미안이라 지금 며씨인가요?
19. 그는 성격이 참 문안하다.
20. 대학을 나왔다고 맞춤법 잘 알거라는 고정간염도 버려요.

Corrected versions are as follows:

1. 카펫이랑 이불이 더럽고 머리카락이 계속 나와서 우리가 청소 새로 다 했어요.
2. 숙소가 많이 낙후되어 있고 엘레베이터 없고 4층이라 짐 많으면 꽤 고생합니다.
3. 화장실 많이 낡았습니다.
4. 그런데 이불들이 너무 머리카락 붙어 있고 얼룩 있어 그래요.
5. 건물도 오래되었음.
6. 된소리를 많이 쓰고 모음을 바꾸는 게 규칙이다.
7. 맨날 할아버지랑 놀고 싶다.
8. 희한하게 며칠 지나면 금세 잊어요.
9. 감기 나으세요.
10. 뒷좌석 벨트 꼭 매세요.
11. 요즘 어이없는 사고가 많대요.
12. 산불이 걷잡을 수 없이 번졌다.
13. 남아일언 중천금이다.
14. 밥도 안 주고 잠도 안 재우는 건 가혹행위 아니냐?
15. 나한테 이래라저래라 하지 마.

---

[12] Sources for some of the examples: 1–5: https://gametubess.tistory.com/107, accessed November 17, 2024; 8–11: https://news.sbs.co.kr/news/endPage.do?news_id=N1003209822, accessed February 2, 2025; 13–19: https://www.100news.kr/8215, accessed February 2, 2025.

16. 이 옷 외숙모가 사주셨어.
17. 이 돈이 내가 낼 수 있는 맥시멈이야.
18. 다름이 아니라 지금 몇 시인가요?
19. 그는 성격이 참 무난하다.
20. 대학을 나왔다고 맞춤법 잘 알 거라는 고정관념도 버려요.

Translations as follows:

1. The carpets and blankets were dirty and hair kept coming out, so we cleaned them all.
2. The accommodations are very old, there is no elevator, and it is on the 4th floor, so if you have a lot of luggage, it is quite difficult.
3. The bathroom is very old.
4. But the blankets have so many hairs and stains.
5. The building is also old.
6. The rule is to use a lot of tense consonants and change vowels.
7. I want to play with grandpa all the time.
8. Strangely, I forget everything quickly after a few days.
9. Recover from your cold.
10. Make sure to fasten your seat belt in the back seat.
11. There have been a lot of ridiculous accidents lately.
12. The forest fires are spreading out of control.
13. A man's word is worth a thousand pieces of gold.
14. Isn't it cruel to not give me food or let me sleep?
15. Don't tell me what to do.
16. My aunt bought me this outfit.
17. This is the most I can afford.
18. What time is it now?
19. He has a really nice personality.
20. Let's get rid of the stereotype that college graduates are good at spelling.

We used the following prompt, in English:

```
The following 20 sentences may contain misspelled words.  Correct
the spelling errors in the sentences to the standard Korean
spelling. The original sentence includes many spelling errors with
similar pronunciation with the standard one. In your response, plea
retain the numbers (1., 2. etc.) with each sentence and put your
entire set of corrected sentences within an <OUTPUT></OUTPUT>
tag:
```

We had also tried prompting in Korean: this produced mixed, but generally worse outcomes.

One possible concern is that the Korean misspellings may be *more* different from the correctly spelled versions than is the case for English. To test this, we computed a simple Levenshtein distance between the pairs of misspelled and corrected sentences at the character level, normalizing by the lengths of the correctly spelled sentences. In the case of Korean we first decomposed the syllables into *jamo* before computing the edit distance. The normalized distances were almost identical, 0.16 in the case of English, and 0.17 in the case of Korean. This suggests that the performance differences between English and Korean cannot be attributed to the Korean set being harder.

Chapter 5
Gesture complexity

5.1 Prelude

In the final episode of season 4 of the American television series Breaking Bad (no spoilers), the character Hector Salamanca, played by actor Mark Margolis, makes use of a physical alphabet/letter board to construct a message in collaboration with a caregiver, who serves as a communication partner. Hector, a recurring character in the series, is confined to a wheelchair and can neither speak nor write with a pen nor type with a conventional keyboard. Instead, he communicates by means of a single call bell, similar to what one finds at reception desks of hotels to request attention, which he can tap with an index finger to produce a chime. For Hector, among other things, the bell serves as a binary switch for indicating yes (1 chime) or no (0 chimes). When presented with an option to possibly select, a chime indicates: yes, select the option. No chime within some suitable time span indicates that the presented option was not selected, at which point subsequent options can be presented. To communicate with a letter board, such as that used with Hector (reproduced in Figure 5.1 and discussed in more detail below), the communication partner presents each row in the grid as an option by saying the first letter in the row. When a row of letters is selected by a chime, the communication partner then presents each letter in the row by speaking it. Here a chime selects the spoken letter, which is added to the message and the process starts again. This general approach of first selecting the row, then the character in the row is known as row/column scanning (Beukelman & Mirenda, 1998); and the involvement of a human communication partner in the process is known as partner-assisted scanning, as mentioned in Chapter 1 in the context of Jean-Dominique Bauby's writing method.[1]

In the episode, the caregiver establishes that Hector wants to create a message using the board by receiving an affirmative chime after asking whether

---

[1] In Bauby's case, they did not use the row/column approach but instead linear scanning—one letter at a time—which we will discuss later in the chapter.

© The Author(s), under exclusive license to Springer Nature Switzerland AG 2025
B. Roark et al., *Tools of the Scribe*,
https://doi.org/10.1007/978-3-032-00831-2_5

A	b	c	d	1	2
E	f	g	h	3	4
I	j	k	l	m	n
O	p	q	r	s	t
U	v	w	x	y	z
5	6	7	8	9	0

Fig. 5.1 An AEIOU alphabet (or letter) board, as used in Breaking Bad.

he would like to use it. She retrieves a transparent letter board (visible from both sides) with the 6-by-6 grid shown in Figure 5.1. To construct the message, the caregiver says "A, E, I", at which point the bell chimes (ding!), then "I, j, k, l, m, n" (ding!). The caregiver then writes down the first letter of the message: 'n'. Then: "A, E" (ding!), "E" (ding!), i.e., the next letter is 'e'. In such a way, they jointly successfully create the message: "needdea". When the caregiver probes for the next letter ("A, E, I, O, U"), Hector does not chime the bell, an indication that the message is complete. "I don't think we're finished Mr. Salamanca," the caregiver states, not understanding the intended message. She then tries again to elicit another letter ("A, E, I, O, U"), again no response. She correctly segments the letter string into "need dea" but feels that the second word is incomplete: "Finish the word Mr. Salamanca. Is it 'dear'? Is it 'deal'? Honey, 'dea' ain't a word, help me out here." Although we do not see it, the caregiver apparently eventually understands that 'dea' represents the initialism D.E.A.

This brief fictional interaction is peripheral to the unfolding narrative of the drama—hence the lack of spoilers—but contains many elements that are central to the topic of this book as a whole and specifically to this chapter. Despite apparently relatively severe mobility- and speech-impacting physical impairments, the individual is able to slowly but effectively dictate a written message to the caregiver. Importantly, relatively little movement was required from the individual to produce the message, compared to other possible means of producing it, e.g., dictation by speech, handwriting, or direct selection of letters via pointing. This was accomplished by the participants jointly establishing a binary code—a sequence of zeros and ones—for each letter on the grid, and an algorithm for eliciting the binary code from the individual via the bell, which we will explicitly illustrate shortly. Additionally, the dialogue between the communication partner and the individual that resulted in the message demonstrates the potential role of intelligence in such co-construction scenarios, either human intelligence or machine intelligence (for messages constructed without a human communication partner). Let us briefly elaborate on each of these aspects of the scenario before diving more deeply into these general topics over the course of this chapter and some of the next.

The fastest and most common means for communicating such a message to a caregiver in this scenario would have been to speak it. Speech, of course, requires highly complex temporal coordination of articulators (tongue, lips, glottis, etc.) to produce an intelligible linguistic message, which can be challenging for many. Other modalities are generally available apart from speech, such as handwriting. Handwriting, like speech, involves a fairly complex sequence of coordinated movements (of hands rather than articulators), placing it beyond the abilities of some individuals. Presented with the grid in Figure 5.1, an individual might directly select the desired letters by pointing. Pointing can happen with a manual pointing gesture, with physical pointers attached elsewhere on the body, e.g., to the head, or even by just looking, provided the direction of eyegaze can be precisely tracked. For this particular individual, apparently such direct selection methods were also inaccessible, hence the use of *indirect* selection via row/column scanning was used.

Note that this kind of row/column scanning can be (and is) automated in soft (virtual) keyboards on computers or mobile devices (famously used by Steven Hawking, as mentioned in Chapter 1), allowing individuals to create messages without the assistance of a communication partner. This is known as computer-assisted scanning. A switch of some kind (i.e., a selection mechanism) is attached to the device, and activation of the switch (e.g., pressing of a button) serves the same purpose as a bell chime in the above scene, selecting the row of letters or specific letter. This is a writing method that we will come back to later in the chapter.

The specific grid used (Figure 5.1) has some pros and cons for this kind of indirect text entry. It is a known style of letter board, so the creators of the show did their homework. On the pro side, it is in alphabetic order, so that it is easy to find the target letter on the grid, even for those without prior experience using such an approach. Second, vowels, which are relatively frequent as a class of letters, occupy the preferred locations in each row. Every letter requires two chimes to select in this algorithm, but the letters vary in the number of unselected options that are presented before reaching the target item. Each option that goes unselected takes time to present and wait for a selecting chime, hence one can measure how long it takes to select each letter by how many such options must be presented, which is determined by its position in the grid. The higher the row, the earlier that row is selected, and the further a letter is to the left in its row, the earlier one selects that letter in the row. So the letter 'A', located in the top left corner of the grid, is the fastest to select of all the letters in the grid (two quick chimes), followed by 'E' and 'b' (both requiring just one extra presented option), and so forth. Thus vowels being in the first column means that they are relatively quickly selected. On the con side, however, in this particular grid common letters in English such as 'n', 's' and 't' require more time to select than very infrequent letters such as 'j' and 'q'. One could rearrange the letters in the grid so that common letters occur in the upper left-hand corner (faster) and less common letters occur in the bottom right-hand corner (slower). With such an arrangement, however, the individual cannot rely on alphabetic order to determine where the target letter resides, but instead must visually locate their target letter on the grid, which may be challenging for those unfamiliar with the method or with difficulty visually scanning. Still, these kinds of optimizations can reduce the typical amount of waiting involved for scanning to proceed, which will speed composition. Speeding up writing is the topic of the next chapter, so we will examine such optimizations in more detail there.

Another way to demonstrate that the upper left-hand corner is preferred versus the lower right-hand corner in row/column scanning is to show that the binary codes assigned to letters are shorter in the upper left-hand corner than in the lower right-hand corner. If one takes Hector's lack of chime as a 0 and his chime as a 1, to select the letter 'A' requires a 1 (to select the row) and another 1 (to select the letter in the row), yielding a binary code of 11 for 'A'. In contrast, to select the letter 'E' requires a 0 (to not select the first row), then a 1 (to select the second row), and another 1 (to select the letter in the row), yielding a code of 011. The code to select an 'n' (which starts Hector's message) is 001000001 (sixth letter in the third row), which is pretty long (i.e., slow to enter) for a very common letter in English. Whatever the input mechanism, longer codes require more work to input, hence this can be seen as contributing to the complexity of the gestures required to produce the text. As mentioned earlier, one can design a grid (and/or other methods) to reduce the *expected* code length as one means for reducing the complexity of the gesture required for input.[2] The idea of a 'code' to specify text will arise at multiple points in the chapter.

Finally, the caregiver engages in some co-construction with Hector, trying to guess a completion of the second word (dear? deal?) based on the first three letters of the word. This kind of completion of words, phrases or even sentences can speed message construction greatly.[3] In a quote we presented in Chapter 1, Jean-Dominique Bauby bemoaned the reluctance of his communication partners to engage in such speculation. Providing such completion options can also be automated, and hence becomes a legitimate

---

[2] One can establish expected code length by measuring code length for some 'typical' sample of sentences, where 'typical' will depend on the language, the individual, and the communication context—basically text that is representative of what is likely to be input.

[3] In fact, as we saw in the scene, the method requires the communication partner to guess where the word boundaries are, since there is no space bar in the grid. So the intelligent interaction is not just to speed things up, but in this case is required to make it work at all. True segmentation ambiguities (such as `theremix`, which could be `there mix` or `the remix`) presumably do arise in natural interactions.

part of the design of text entry technology, as the means for simplifying the steps required to produce the desired text. However, this comes with some downsides. First, such an interaction requires that the user attends to and responds to suggested completions, which represents an increase in the gesture complexity required from the user. Sometimes the suggested completions are useful and are selected, and sometimes they are not useful and are ignored, but the process of attending to a suggestion and then selecting or ignoring is both effortful and time consuming. Whether on balance that leads to easier text entry depends on both the quality of the suggestions and how much effort is required by the user to participate in the interaction relative to other writing gestures.

An important second downside of relying on completion options during text entry is the potential risk of altering the intended message. There is often an important role for intelligence (human or machine) to assist in message construction, yet with many potential pitfalls around message biasing from either human or AI communication partners. Dictation to scribes has been around for a very long time, something that in its most direct application leaves little question about message origination, i.e., who was the author. In the scenario at the start of the chapter, Hector was able to produce his desired message exactly as intended and it was (eventually) understood. The longer the predicted completion, however, the more chance that the communication partner influences the message content, through rewording or even embellishment of the original intended message. When message creation requires so much work, the potential for acceptance of 'good enough' versions lacking fidelity to the author's intent can be high. We will return to such questions, especially when discussing such uses of large language models.

In this chapter we will examine the complexity of gestures required to write, from handwriting to modern text entry methods, as determinants of method accessibility. This complexity may derive from the complexity of movements required to produce the text, such as hand motions involved in handwriting or touch typing, or precisely pointing to a target item with a mouse or via eyegaze. It may also derive from lengthy codes to specify a letter as described above, when methods requiring complex motor skills such as speech, handwriting, pointing or touch typing are not accessible. Such specifications need to be input somehow, and generally the longer the code the more effort required to input them. Additionally, the demands of giving attention to presented options or visually scanning to find target options can also accrue to the effort and must be taken into account when assessing the complexity of required gestures.

5.2 Baseline writing methods

5.2.1 Spoken dictation

As we have repeatedly taken pains to point out, language is distinct from writing, yet they are often confounded. Some languages, in fact, are conventionally written in different scripts by different speakers, such as Punjabi, which in Pakistan is commonly written in the Shahmukhi writing system using a Perso-Arabic script while in India it is commonly written in the Gurmukhi writing system using a Brahmic script. Those who are literate only in Shahmukhi can, of course, perfectly understand the speech of those who are literate only in Gurmukhi and vice versa, since they speak the same language—they just cannot read each other's written messages. The transcription of language via a particular writing system is one of potentially many conventions for representing the language in writing, and the key requirement for successful communication via that transcription is that the message is understood by the recipient, whatever the modality. In fact, even the author of the message need not be able to read it, if there is some method for creating the transcription of their message, such as dictation, though that may leave the author unable to verify that the written text accurately reflects their message.

Typically the most direct method to use language is to speak (or sign), and hence methods to convert spoken or signed language to writing are often the most direct way to write. For example, a person can take dictation from the author, i.e., write down their intended message; or automatic speech recognition (ASR, also known as speech-to-text, STT) can be used to transcribe the author's speech.[4] The use of an amanuensis to take dictation has a long history. To take an example from the ancient world, Bagnall & Cribiore (2006) point out in their study of women's letters from ancient Egypt that those most likely to be literate were also most likely to be able to afford a clerk/secretary to assist with correspondence, hence dictation was common even when the author was literate (which itself was relatively uncommon). In the corpus of Greek and Coptic letters covered in that book, there is extensive evidence of dictation.[5] They note that requests to the assistant could vary from full dictation of the exact contents of the correspondence (which would often result in what they describe as a more "oral" style versus polished professional writing) to general instructions to write some particular content to the recipient, leaving the specific wording up to the assistant. In modern

---

[4] Since signed languages are not generally written, we will here focus on spoken language. Should writing systems become widely used for signed languages, most of the points being made here would also apply.

[5] This forces the authors to address the question of whether a letter represents the actual words of the author or is something wholly written by the assistant, perhaps at the request of the author.

times, such dictation scenarios also existed in the early days of typewriting, when the use of such machines was typically the province of trained specialists, and human dictation likely continues to be used in specific circumstances to the present day.

One might argue that giving dictation is beyond the scope of this book, focused as it is on "tools of the scribe", since the scribe in this scenario is the person taking dictation rather than giving it. However, this underestimates the degree to which such dictation scenarios may involve a dialogue between the scribe and the author, who may collaborate in message construction (much like the collaboration with the caregiver presented at the beginning of the chapter), such as asking for clarification or suggesting a repair. Further, with the increasing prevalence of artificial intelligence, such collaborations for message construction may often occur between a human author and a machine communication partner. In that setting, the machine is the tool and the human speaker is the scribe.

A key reason to use speech for writing via dictation is speed. Pellegrino et al. (2011) report the number of spoken syllables per second in typical speech across a number of diverse languages, and find values in the range of 5-8 syllables per second in all of the languages they examined.[6] The rate of English in that paper was reported to be around 6 syllables per second, which would seem to roughly correspond to the commonly quoted 150 spoken words per minute that is often used to compare with (slower) handwriting and typing speeds. One merely has to try transcribing spoken language in real time to discover how difficult it is to keep up with typical speech with standard writing methods. This is why stenographers use shorthand, something we will discuss in the next chapter.

Other reasons to use speech for writing include not needing to use one's hands to produce the text. Indeed, Rayner et al. (2006) motivate a spoken dialogue system developed by NASA for use on the International Space Station, by a key consideration of enabling computer access when the hands were otherwise occupied. Injuries or lack of manual motor control are other reasons that speech may be a preferred modality. Also, as stated above, speech does not require literacy, so can be used for writing by those who cannot write themselves.

On the flip side, speech is harder to keep private than other writing methods, since others besides the amanuensis can hear it. Imagine a large cafe near a college campus on a weekday afternoon near the end of term, when it is full of students writing on their laptops, diligently working on term papers (or perhaps not so diligently texting with friends). Using speech input for this writing would additionally share everything being written with others in the cafe within hearing distance. Any writing method has privacy

---

[6] Handwriting and typing speeds are often quoted in words per minute, but the unit of 'word' differs significantly across languages, while syllables are much more comparable cross-linguistically.

concerns—screens can be read, text can be intercepted, etc.—but speech has this additional audible dimension.

Additionally, the students' term papers (or even text messages) are not generally simply written in a single linear pass from start to finish, rather they are edited and modified as part of composition. Spoken dialogue with either a human or machine as part of the editing process is certainly possible and can be very effective, but direct access to the written text itself makes possible certain modifications that are less directly made via speech. One common mode for using speech to write is as a first pass, the result of which is then returned to the author as text to review and edit—this is true of both dictation to a human and modern automatic speech-to-text systems.[7]

There exist scenarios where speech is the best option due to the inaccessibility of other conventional writing methods. For example, severe motor impairments can make handwriting or typing very slow or even impossible. Automatic speech-to-text (STT) improvements have resulted in highly usable systems for most individuals in languages for which such systems are available, which includes most widely spoken languages. The speed of speech makes it attractive relative to other potential text entry methods for those with severe motor impairments, such as the use of letter boards as presented at the start of this chapter. However, many individuals with severe motor impairments that would recommend using speech for writing also have corresponding speech impairments that impact intelligibility. Speech motor impairments, collectively called dysarthria, can cause slurred or otherwise poorly intelligible speech, and are commonly found in congenital conditions such as cerebral palsy (CP) as well as in degenerative disorders such as amyotrophic lateral sclerosis (ALS) or Parkinson's Disease (PD).

Severe dysarthria can impact intelligibility for human listeners as well as for automatic speech-to-text systems, impacting the accuracy of transcription whatever the method. Even so, given the speed of speech, this remains an attractive option for individuals, and Koch Fager et al. (2019) point out that many individuals with severe speech impairments want to use speech, hence advocate for development of systems to improve transcription of dysarthric speech. As an example of methods that can improve the accuracy of transcription, they highlight a method originally published in Hosom et al. (2010) whereby the initial letter of a word is typed/selected and the word is spoken, thus reducing the ambiguity faced by speech recognition and also reducing the effort required versus typing the whole word. This approach was integrated with a word prediction engine, thus further reducing the effort of typing. Individuals with dysarthria for whom typing gestures are difficult (or fatiguing) could find this strategy effective relative to other writing options.

---

[7] We defer to later in the book—see Section 6.4.5—discussions of using AI in the dictation mode that involves giving instructions about what should be written rather than straightforward transcription of the speech, since this is not purely a speech related topic. Needless to say this is germane to the scenario above of students writing term papers in a cafe.

Koch Fager et al. (2019) also present cases where individuals want to use their speech to communicate face-to-face with people. Their friends and family are familiar enough with their speech that they can understand it, and this is a fast and natural way for them to communicate. To communicate with people who are unfamiliar with their speech, however, they must rely on synthesized speech that is more intelligible than their own speech. To do this, they must use some writing method to create text, which is then spoken by a text-to-speech system. The writing methods available to them are often slow relative to speech, so using their own speech—even to produce synthesized speech—remains an attractive option. However, the characteristics of their speech that impact intelligibility for the people they want to speak to also makes automatic speech-to-text challenging.

Methods to make automatic speech-to-text accurate enough to effectively transcribe dysarthric speech for such systems—or for just producing text in non-face-to-face scenarios—continues to be an area of active research. Much of this work focuses on personalized training of speech recognition systems, meaning that the individual user provides speech samples and correct transcriptions to the speech recognition system to include in its training. This contrasts with speaker-independent systems that have no prior knowledge about the particular speaker. As an example of this approach, Green et al. (2021) showed that personalized models heavily outperformed speaker-independent models, and in fact outperformed human listeners overall and most particularly for the most impaired speakers, across a diverse set of participants with a range of speech-impacting conditions, including ALS, CP and PD. Pursuing such an approach requires providing training examples up front, hence imposes some configuration effort on the part of the user—hopefully with some system performance payoff. Note, however, that just training the system once may not be sufficient. Tomanek et al. (2023) showed that for degenerative disorders such as ALS, speech recognition performance degraded over time for both speaker independent and personalized systems, but that inclusion of recent examples for training personalized models mitigated this degradation. In other words, such systems will benefit from on-going personalization.

Even with improved performance, however, there remain questions of how exactly to integrate the technology into the communication options available to the individual. Cave (2024) examined how individuals with ALS actually used speech recognition in their daily lives—including with adaptations specifically tailored to dysarthric speech—which highlighted many remaining challenges in tailoring the technology to actual use cases. The potential of such systems to improve the writing options for individuals with severe motor impairments is real, but too often unfulfilled in available systems.

With respect to the topic of this chapter, sometimes the easiest, most direct way for an individual to write something is to speak it and rely on an external (human or machine) transcription of the speech—a scenario that has been actively used for millennia. While this should be considered a baseline writing method, in aggregate the gestures required to write in this manner are relatively complex. Speech production itself is complex, making this approach inaccessible for some. Even for those for whom speaking is easy, the overall process—including reviewing and editing transcripts, however they are produced—may require more work than other methods of directly writing. Further, the editing process typically includes other methods of writing, which can more directly manipulate the text. For short messages, however, such as search queries or brief text messages, spoken input that is converted to text is both common and effective.

5.2.2 Handwriting

While speech is a natural modality, the true *baseline* writing technology—used across languages and centuries to the present day—is to use some sort of implement held in the hand to apply marks to a surface. In much of history, and universally in the modern world, the canonical instance of this is pen and paper, whereby markings (ink) are inscribed on a medium (paper) via manual movements of an instrument (pen). Modulo improvements in types of instruments, inscriptions and media over the years, this mechanical technology has been around (by definition) since the dawn of history. While perhaps not the most common writing technology anymore, it has its role to this day, i.e., it has not been wholly displaced by newer technologies.

Writing a page of text with this baseline writing technology involves a sequence of manual movements (gestures) with the pen that typically leave markings in lines on the paper, which ideally other speakers of the language can read. Handwriting is a complex, learned skill which is generally acquired much later than speaking and mastered later than reading. Expertise can be further extended to a mastery of the art form of calligraphy, and for most of recorded history, skill in handwriting alone could form the basis of a profession.

Like speech, the movements required for handwriting can be difficult in the face of physical impairment. Even in favorable circumstances, the sequence of manual gestures involved in handwriting can be physically taxing; for example, students who hand write essays as part of a lengthy end-of-term exam may experience discomfort in their hands by the end. Calluses can form on the hands of frequent hand-writers where the pen typically touches the fingers. What became known in modern times as Repetitive Strain Injury has been attributed to excessive handwriting for hundreds of years (Bloemsaat, 2006). It is certainly not effortless, even for the most practiced hand. Some people struggle to produce legible handwriting, and for others this is not an accessible technology at all, such as when disabilities make performing the required sequence of manual gestures extremely difficult or even impossible. An injury that impacts the hands—e.g., if the hands must be placed in rigid casts—can have an effect on handwriting for anyone, at best making the resulting writing more difficult to produce and less legible; and at worst, painful, extremely effortful or even impossible. Many of the motor impairments discussed in Section 5.2.1 also impact handwriting legibility, making handwriting equally inaccessible.

However, in order for alternative writing technologies to displace handwriting as the writing method of choice, there must be some kind of advantage to the alternative over the baseline. These advantages may not be universal, rather tied to specific use scenarios that make the alternative preferable. In some such scenarios, the advantage is very easy to see. For example, the movable-type printing press involved "origination" of the document (i.e., typesetting) just once, followed by multiplication (or reproduction) of the typeset document many times. In scenarios where many copies were to be produced—such as a pamphlet in the 19th century—the extra effort required to typeset the document versus writing it by hand was rewarded by only having to do it once, however many copies were printed. If it took 100 times as long to typeset the pamphlet as it did to write it by hand, but 1000 copies of the pamphlet were being produced, then it would have been (roughly) 10 times faster to produce the 1000 pamphlets by printing press. This was clearly a speed advantage over handwriting, but perhaps also an advantage in terms of legibility and compactness of the printed versus handwritten text and possibly the ease with which errors could be corrected, all of which made this technology clearly superior for this use case. Even with these advantages, however, it would typically have made little sense to typeset a document if few reproductions were to be made.

For use scenarios when only a single version of the document is being produced, such as writing a letter or some other form of correspondence, there are also some situations where alternative writing methods are clearly superior to handwriting. In particular, when disabilities make handwriting impossible, but pushing/pressing/selecting one or more keys is possible for the individual, then a writing technology based on such a gesture will be better— not necessarily *good*, but certainly better than handwriting, since the latter cannot be accomplished at all. Hence, while handwriting exists for all written languages, thus is as accessible in one language as in another language, there are many individuals worldwide for whom handwriting is not an accessible writing technology. For some of these, typing on a typewriter or similar device, does represent an improvement in accessibility over handwriting, largely due to differences in the kinds of gestures required to create the transcription.

Individuals for whom the baseline writing technology—handwriting—has been a largely acceptable method will typically require some further reason to use another technology. Today, inputting text from a keyboard on a laptop or via a soft (virtual) keyboard on a mobile device provides machine-readable text, with all of the benefits that are derived from that, such as cutting, pasting, emailing, storing, typesetting, translating, general computing, etc. It is possible to derive machine readable text from handwriting through the use of handwriting recognition systems, and for some people and in some writing systems this can be a technology of high utility. As with speech, such technology has its analogue in office technology of the 20th century: the typist. Typists as a professional class were responsible for taking handwritten documents and converting them to typed documents. They may have created the handwritten documents themselves while taking dictation, or the documents could have been handwritten by the authors. Like taking dictation, almost surely such practices, while less frequent, still do occur—e.g., an administrative assistant converts a handwritten note to an email. Alternatively, one might use machine intelligence to correctly identify the intended message and convert it to a digital representation.

Mechanical precursors to machine-readable text entry—in particular the typewriter—did not have all the above-listed benefits of producing machine readable text, yet many such devices were invented and some eventually became ubiquitous. The residual influence of typewriter design on text entry and the role of accessibility in motivating such mechanical device development make that history very relevant to this chapter's topic. Several obscure long-forgotten mechanical writing methods usefully illustrate the tradeoffs between accessibility, flexibility and speed that persist in text entry methods today.

5.3 Mechanical writing

We discussed typewriters in Chapter 3 from the perspective of script complexity—specifically issues in typewriter development for non-linear scripts. Here we will cover the topic from the perspective of gesture complexity and why such efforts were expended to develop a mechanical alternative to writing by hand. Along the way we will examine a few largely forgotten alternatives to what became the de-facto conventional typewriter design.

5.3.1 Why type?

Beginning in the late 19th century and continuing through the 20th century, mechanical typewriters became commonplace in some regions of the world as an alternative writing method to handwriting. By the time the conventional design of the typewriter emerged in the late 1870s, there had been sustained efforts over a century and a half by many individuals to design/create mechanical writing machines, largely in western Europe and the United States—places with languages making use of the Latin script (for French, German, Italian, Danish, English, etc.), which is relatively well suited to the task as discussed in Chapter 3. The generally acknowledged first documentation of a writing machine design came via a patent in England in 1714. We can examine the high level rationale provided in that patent by the inventor,

Henry Mills, for advantages that such a machine was hoped to provide over handwriting. The patent claims:

> "An Artificial Machine or Method for the Impressing or Transcribing of Letters Singly or Progressively one after another, as in Writing, whereby all Writings whatsoever may be Engrossed in Paper or Parchment so Neat and Exact as not to be Distinguished from Print; that the said Machine or Method may be of great Use in Settlements and Publick Recors, the Impression being Deeper and more Lasting than any other Writing, and not to be Erased or Counterfeited without manifest Discovery;"

Unfortunately, the Mills patent contained no specifications for the said writing machine, but regardless of the actual design, we see several claims being made (rightly or wrongly) of advantages of the machine over handwriting:[8] neatness; all the benefits of (i.e., indistinguishable from) print—which would include legibility, compactness and uniformity; longevity of documents; and difficulty of altering or counterfeiting. Adler (1973) points out that advantages claimed by early inventors over handwriting included uniformity, legibility, ease of copying and speed. Compactness, Adler adds later (p. 68), can additionally have a direct cost impact, since paper was heavy, which can make compactly written letters less expensive to send through the post, so this can also accrue as a benefit of typed text.

Additionally, many early efforts to create writing machines were focused on technology for the blind, which typically involved embossing the text instead of or in addition to inking. Indeed, in sorting through the historical record, Adler comes to the conclusion (p. 164) that Pellegrino Turri's machine, built in 1808 to permit a blind woman to correspond, was the first documented use of a mechanical, sequential (letter-by-letter) writing machine, due to the surviving correspondence that it produced, hence claiming the title of the first actual functional typewriter invention. From this, we can add accessibility to the list of potential benefits claimed for typewriters and similar machines.

We can thus boil down the (not necessarily exhaustive) list of potential benefits of typing over handwriting to: those related to looking like printed text (neatness, uniformity, legibility); those related to beneficial properties of the resulting document (compactness, ease of copying, document longevity, difficulty of altering or counterfeiting); and those related to text production (ease, speed and accessibility). In this chapter, we are focused on the latter issues, which often come down to properties of input gestures required to produce the text. Simple, fast actions are generally preferred to complex, slow actions, all else being equal.

We note in passing that a potential benefit of one writing method over another that was not mentioned above is *learnability*, i.e., how hard/easy is it to obtain proficiency in the method. Often this takes the form of the related question, whether it is worth investing the time to learn the new method given

---

[8] In Chapter 3 we saw some scripts for which some of these advantages never accrued from mechanical writing.

what it offers over and above existing methods in which one is already proficient. In Chapter 1 we speculated about the reasons why Jean-Dominique
Bauby used partner-assisted scanning instead of Morse code to communicate
using eye blinks, and noted that one key relative disadvantage of Morse code
is time to learn. Handwriting is probably not the writing method with the
easiest learning curve to achieve proficiency, yet it is the method that most
children are taught first, before keyboarding. Then, generally, subsequent decisions about whether to acquire proficiency in new writing methods will boil
down to the cost of time/effort invested learning the new method versus its
benefits over existing methods. The benefits of keyboarding over handwriting
in the current day are generally significant, but were typically much less so
for mechanical typewriting.

5.3.2 Typewriter design choices

Unlike comprehensive histories of typewriters, such as Adler (1973) or Beeching (1990), we are not particularly interested in exhaustively detailing the
history of invention that ultimately led to the widely used technology; nor in
tracking down and assigning credit to the person who had the stroke of genius
that made it all work. However, we can look at a selection of mostly forgotten
writing machines that were produced during this fertile period of invention,
and how they may have addressed different needs of different users. Much of
our current writing technology—on laptops and mobile phones—is at least
partly based on methods designed for mechanical typewriters, with many of
the same pros and cons of that technology. Common contemporary text entry methods can be slow and/or inaccessible for particular populations for
more-or-less the same reasons that common typewriters were. Interestingly,
some of the machines that did not become widely used had characteristics
that can help us understand some of the key trade-offs that still exist. Before turning to these, however, we will consider the typewriters that became
conventional.

5.3.2.1 QWERTY

The conventional keyboard design that emerged in the late 19th century
and is still widely used today is the so-called QWERTY keyboard, after the
first six letters on the upper row of the keyboard. It involves three rows of
7–10 letters per row, often with an extra row of other characters, typically
digits and/or punctuation. For example, Figure 5.2 shows the keyboard
of a Remington No. 5 typewriter from around 1888, with the familiar grid

Fig. 5.2 Remington No. 5 typewriter, c. 1888, at the Museo della Macchina da Scrivere, Milano. Source: own work.

of letters and other symbols.[9] You can see that this is not the QWERTY layout but rather the AZERTY layout that became conventional in France and Belgium. In the AZERTY layout, compared to QWERTY, the keys for A and Q are swapped, as are Z and W, and the placement of the M key is in the middle rather than bottom row. There were also QWERTZ (Germany, Austria and Switzerland) and QZERTY (Italy) layouts with similarly minor key placement changes, and all of these have persisted into the modern era as conventional regional keyboard layouts. The only potentially meaningful change of any of the above layouts is the Y/Z alternation in the QWERTZ layout, which places Z nearer the index finger during touch typing. One might be tempted to ascribe these alternations to statistical characteristics of the target languages—i.e., some ergonomic, efficiency or mechanical optimization that takes word frequency in the language and orthography into account— but, even considering the Y/Z alternation in the QWERTZ layout, these are only minor changes at the periphery of the keyboard, without any obvious language-specific optimization, any more than QWERTY for English.

Why QWERTY and related orderings? Adler (1973) provides two explanations which are now widely known. The first is mechanical: that the arrangement was arrived at through trial and error to avoid type-bar collisions when the arms for two different letters were typed in quick succession. Each type was mounted on the end of a type-bar that would carry the type from its position in the carriage to the striking point where the impression was made. If multiple type-bars were activated simultaneously, i.e., multiple keys pressed at the same time, they may arrive at the same point at the same time, caus-

---

[9] You will notice from captions of a number of the typewriter images in this chapter, the typewriter museum in Milan is one of our favorite fieldtrips, highly recommended. The collection is detailed in Di Donato (2019).

ing a collision and possibly a jam (i.e., failure of the type-bars to retract). Even if the keys were not pressed simultaneously but rather just in relatively quick succession, the possibility of collision/jam was real, particularly if the type-bars associated with the keys originated from nearby resting locations. If typing were slow, however, each type-bar would return to its resting location before the next one was activated, hence largely avoiding this situation. Some have gone so far as to claim that the keyboard was deliberately made inefficient so as to slow typing and thus reduce the frequency of type-bar collisions. The second explanation offered by Adler (1973) is that all of the letters to type the brand name TYPE WRITER fall on a single row of keys, making it easy and fast for salespeople to type that word during sales demos. Apparently the European models didn't require such fancy demos, since none of the layout variants mentioned above other than QWERTY have this latter property.

These two explanations, however, do not do justice to all of the influences on the inventor, Christopher Latham Sholes, during the evolution of the keyboard layout that resulted in QWERTY. Adler (1973) points out that the original keyboard was in alphabetic order, remnants of which can be seen in the middle row of letters in Figure 5.2 from D to M, which fall in alphabetic order (skipping over vowel letters). But Sholes requested a list of common English letter pairs from a cohort as part of the effort to optimize the mechanical operation (Sproat, 2010, Section 6.2.1.1). This was presumably used to improve avoidance of likely type-bar clashes by ensuring that the type-bars of each letter in a common pair were far apart. Sproat (2010) demonstrates that one cannot easily judge type-bar proximity from keyboard proximity—in fact, relatively close keys on the keyboard, such as T and H, fall on opposite sides of the type-bar basket. Keyboard position did determine type-bar placement, however there were complexities in the mapping, clearly explained in Sproat (2010), that would have made choices about relative key placement of common pairs relatively subtle. With the advent of touch typing,[10] more careful and systematic keyboard optimizations based on language statistics—which we discuss below—ultimately failed to yield much improvement in typing speed over QWERTY, and metric-based assessments in Sproat (2010) point out that, for many ways of assessing keyboard layout efficiency, QWERTY yields surprisingly good scores. This suggests that deliberate keyboard inefficiency as a strategy to reduce type-bar clashes was unlikely (or very poorly done). Whether fairly decent QWERTY keyboard efficiency is due to some inspired keyboard location choices being made somewhat independently of type-bar clash avoidance is unclear but very possible. Further, if type-bar clashes were the whole story, one might have expected more changes to the regional (French, German, Italian) QWERTY variants than were actu-

---

[10] Sproat (2010) further argues that it is unsurprising that very frequent letters are missing from the "home" middle row in QWERTY since having them there is specifically a touch typing optimization, and touch typing as a practice did not exist until well after the device was invented.

Fig. 5.3 Smith Premier typewriter, c. 1889, at the Museo della Macchina da Scrivere, Milano. Source: own work.

Fig. 5.4 Caligraph typewriter, c. 1882, at the Museo della Macchina da Scrivere, Milano. Source: own work.

ally carried out. In any case, the narrative that the keyboard was designed solely through brute force trial and error to reduce type-bar clashes is woefully incomplete.

The Remington had some features that we take for granted now, such as the shift key (labeled in Figure 5.2 with the French *majuscules* for 'uppercase'), thus cutting down on the number of required keys. In contrast, the Smith Premier typewriter, which was a contemporary of the Remington, duplicated the alphabet in both upper and lowercase rows, as shown in Figure 5.3, which is known as a 'double' keyboard. This leads to a rather large matrix of character locations, and presumably a more effortful manual operation to move from the lower keys to the upper keys versus simply selecting the shift key. Still, the complexity of even this large keyboard seems modest compared to the Caligraph keyboard, also from the 1880s, shown in Figure 5.4. Here many of the patterns of the QWERTY keyboard are found, such as the lowercase letters *asdfgh* found in the middle row in sequence. However, this uses what was known as a 'full' keyboard rather than the 'double' keyboard of the Smith Premier: the uppercase letters are now on either side, without the same kind of systematic spatial relationships found in the 'double' layout. For example, lowercase 'c' and 'k' are adjacent on the right side of the middle row, but their uppercase counterparts are far away from each other—'C' in the lower left and 'K' in the upper right. This would likely increase the amount of visual scanning required to find the uppercase letter associated with a particular lowercase letter position. Even so, according to Adler (1973), this was a commercially viable model for ten years before they changed the design of the keyboard.

Allowing for a shift gesture to make the full uppercase/lowercase inventory available with half the keys is a clear win for speed, which may partly explain why the shift key approach became conventional. Other features of those initial Remingtons, however, were changed over time—most notably these initial typewriters were designed so that the type-bar stroked upwards to impress on the paper, hence the results were not immediately visable to the typist. Down-stroke (hence visible) designs eventually became the standard. But with respect to the keyboard choice, the QWERTY and related layouts have remained the standard from that time forward.

Not all typewriters of that era, however, were so cavalier about fitting the keyboard to the language. The Blickensderfer typewriter, shown in Figure 5.5 made use of three rows of letters, just like the Remington, but used an alternative arrangement of letters across this matrix. According to published instructions for using these typewriters (Blickensderfer Manufacturing Company, 1908), this arrangement was known as the *scientific* keyboard, though it is more commonly known as DHIATENSOR after the keys along the bottom row. If you reshuffle these letters, you can get ETAOINSHRD, the first 10 letters of ETAOIN SHRDLU, a nonsense phrase popularized in the printing industry, representing the most common 12 letters in English in frequency order. The phrase was popularized in printing because the ubiquitous Linotype keyboard—presented in Figure 5.6 as shown in a 19th century manual (McCall, 1898)—has the letters in this frequency order in columns of both lower and uppercase regions. Operators wanting to produce a test slug from the Linotype machine would just hit the buttons in sequence (much as one might type QWERTY on a typewriter just to fill space) and examine the result, thus producing the nonsense phrase.

Fig. 5.5 Blickensderfer typewriter, c. 1893, at the Museo della Macchina da Scrivere, Milano. Source: own work.

Fig. 5.6 Linotype keyboard from McCall (1898)

The layouts of the Blickensderfer and Linotype keyboards in the late 19th century demonstrated an early recognition that the most frequent patterns in a written language should be easier/faster to produce in order to facilitate fast writing. By placing the most frequent letters on the 'home' row of the Blickensderfer typewriter, the designer intended to make access to those letters easier than access to less frequent letters, which would presumably lead to faster and/or easier typing. Such modest optimizations, however, did not prevail, as the QWERTY and closely related layouts (AZERTY, QWERTZ, etc.)

became standard for the Latin script and remain so to this day. More sophisticated later efforts, such as the Dvorak layout (Dvorak & Dealey, 1936), which analyzed letter "digraphs" (pairs of letters, what are also called letter *bigrams*) as part of designing their character distribution over the conventional three row typewriter.[11] While a small but passionate cohort prefers to use the Dvorak layout (so that mobile phone virtual keyboard software will typically support such a layout), usage is dwarfed by those still making use of QWERTY. One reason for this is that claims to increase speed and reduce errors with such a keyboard are controversial. For example, Strong (1956) argued (and demonstrated via controlled experiments) that both Dvorak and QWERTY keyboards achieved fast typing speeds when the typists were trained in fast typing techniques, hence switching was unnecessary, rather improved training was required. In any case, it can be argued that optimizing the character layout of the relatively modestly-sized keyboard required for the Latin script to reduce gesture complexity during typing has a relatively modest impact on typing speed compared to learning touch-typing methods, hence may be judged to be generally insufficient to warrant switching layouts.

Why did QWERTY become the standard for typewriter keyboards, so that switching from that requires some special motivation? David (1985) argues that this is due to natural stochastic processes in the collective adoption of a new technology, in the case when specialists are being trained to operate the technology. In the late 19th century, most typewriting occurred in business rather than personal settings, and most of the business use was by trained specialists: typists. The more QWERTY layout keyboard typing jobs there were, the greater the reward for learning to type with that layout. The more trained QWERTY typists, the greater the reward for a company to provide that specific writing technology, in terms of being able to hire trained operators. Relatively small early differences in scale between alternative layouts can lead to eventual complete dominance of one over the other, which is what happened. The ease of changing the placement of characters on the keys of a typewriter, relative to other more complex mechanical design choices, then led competitors to offer QWERTY as an option for their own devices, and so on. In this manner, David (1985) speculates, the "historical accident" of our collective adoption of a non-optimal keyboard layout is explained.

The convergence of typewriters onto certain design choices, such as the use of a shift key to handle casing, rather than the double keyboard, had heavy negative impact for some languages. Mullaney (2017) relates details regarding the fate of the early Thai typewriter designed by McFarland (pp. 45–55), which only had to trim two symbols from the Thai script (a Brahmic abugida script, like Devanagari descended from the Brahmi script) to fit all the required symbols within a double keyboard (i.e., no shift key, but all letters on

---

[11] This sort of collocation analysis was heavily used in optimizing the large character matrices used for Chinese typewriting, such as the Double Pigeon device discussed in Chapter 3. Characters that would frequently cooccur could be placed near each other to minimize operator movement during operation—see Mullaney (2017) for details.

the keyboard, see the Smith Premier in Figure 5.3). This Thai typewriter was then discontinued when Smith moved to the shift key design for a smaller keyboard, which was insufficient to effectively handle Thai. Apparently Thai was too small a market to influence the manufacturers, and nobody stepped up to manufacture an alternative. As detailed extensively in Chapter 3, writing systems using other Brahmic scripts required more draconian changes to symbols to be accommodated by typewriters, due to complexities that preclude full or nearly full coverage with so few keys. Even so, the Thai example illustrates the degree to which these commercial decisions impacted whole classes of people, including those speaking particular languages.

5.3.2.2 One key per letter

Apart from the locations of letters on the keys of the keyboard, the design of the Dvorak and QWERTY keyboards are essentially the same: three rows of letters with one key per letter. One reason such an approach became common was because of the mechanical nature of the device, requiring an impression to be made based on a selection, and the direct correspondence between a single key selection and a type-bar motion is mechanically fairly straightforward (if tricky to effectively engineer). However, there existed writing machines pre-dating the Remington and competitors that did not have a one-to-one correspondence between letters and keys. Investigation of the pros and cons of such machines is instructive, particularly from an accessibility perspective.

Among the simplest of designs that moved away from one-key-one-letter was one of the several writing machines discussed in Adler (1973) that were designed by John Pratt and patented in England in 1866 (Pratt, 1866). In this device, letters were placed on a 6-by-6 grid—similar to the letter board presented in Figure 5.1, enough for the 26 letters plus some extras—and letters were identified by pressing two keys simultaneously, corresponding to the row and column of the letter's location on the grid. The grid is then moved so that the desired character is located at the striking position and an impression is made. In such a way, pressing the two keys simultaneously is simply an alternative way to access letters from a relatively small set, possibly simplifying the hand movements required to access the keys. Other methods used multiple keys to increase the number of symbols that could be produced by the device.

Ten years after inventing his famous eponymous tactile writing system (Braille, 1829), Louis Braille invented another writing method for representing the Latin script that, like his earlier approach, resulted in text that was legible by touch via embossed dots, but which also retained the form of the Latin script so that it was also legible by sight. *Decapoint* (Braille, 1839) was a system for creating what are essentially pixelated representations of letters by placing dots in 10-row columns. The "code" for the letter is the sequence of rows at each column in the overall representation of the letter. Figure 5.7

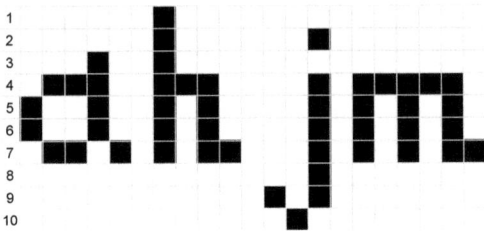

Fig. 5.7 Representations of selected Latin script letters as specified in the Braille (1839) Decapoint system. Source: own work.

presents the Latin script letters *a, h, j, m* using the codes specified in Braille (1839), i.e., for the letter *a*, the code is $(5,6)|(4,7)|(4,7)|(3,4,5,6)|(7)$. This total of 11 dots to represent the letter is in contrast to a maximum of 6 dots for any letter under Braille's better known prior system, so the legibility for the sighted comes at a relatively high cost for coding efficiency.

The *Raphigraphe* machine produced by Pierre-François-Victor Foucault in 1843, provides 10 keys that activate plungers to ink and/or emboss paper in a column, the first practical implementation of Braille's system. Using the correct plungers in sequence results in pixelated forms, in the patterns presented in Figure 5.7. The numbers at the left of the figure show that the "code" for the letter 'a' specifies the pixels that are filled across the five columns of the letter. Note that this seems to be a direct implementation of the ideas that Braille published in 1839, yet in histories of typewriting the focus tends to be on the engineering, so the credit for the Raphigraphe often goes nearly entirely to Foucault. For example, Adler (1973) surprisingly presents Braille's participation as happening post-invention and as being mostly perfunctory, while Richards (1964) omits mention of Braille at all (and mentions the Raphigraphe only in passing).

While the original intent of the approach was to produce embossed dots that were legible both visually and by touch, it is an approach that could have been used in more typical typewriting scenarios. The Raphigraphe device was manufactured and used as a writing machine, especially at the institute for the blind where Braille and Foucault were affiliated, but the general approach was never broadly adopted. One can see one key reason for this when comparing to a QWERTY type-bar style typewriter: to produce a lower-case 'a' with the latter, a single keypress is required; however, to produce the same letter with the Raphigraphe, five separate multi-key presses are required. This would doubtlessly be a more taxing and time-consuming effort—many distinct gestures are required to produce even a single letter. Further, while QWERTY requires knowing where the 'a' is on the keyboard, this device requires remembering the sequence of pixel columns that collectively produce each letter, hence more learning and expertise required from the operator. It

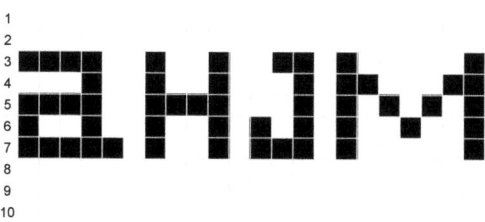

Fig. 5.8 Alternative representations of selected Latin script letters consistent with the Braille (1839) Decapoint system. Source: own work.

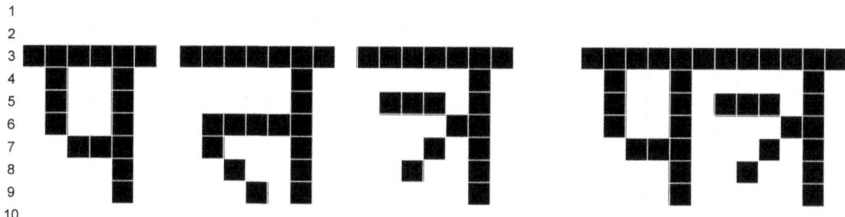

Fig. 5.9 Representations of Devanagari letters, ligatures and words consistent with the Braille (1839) Decapoint system. Source: own work.

does have the virtue of a total of just 10 keys, hence no use of a shift key or moving the hands between rows of keys would be required.

It is the dot/pixel-based conception of the device that leads to such lengthy 'codes' for each letter. Each letter has been decomposed into pixel components, hence producing the letter requires specifying all of the component pixels. This decomposition, however, also provides a tremendous amount of flexibility in designing letters, without changing the device at all. For example, Figure 5.8 provides alternative representations for each of the letters, which are achieved with different letter codes (that we made up) consistent with the same system, still recognizable as the same letters as shown in Figure 5.7. There are some key differences from Braille's codes—in particular, these mix upper and lower cases, and produce the 'double-storey' lowercase 'a'. Note, also, that only rows 3–7 are used in these letters, suggesting that one could design codes to use fewer keys to ease typing and reduce code complexity.

This degree of flexibility in letters that the device can produce is a main source of its complexity as an input device. Instead of just 26 letters in both upper- and lower-case, plus digits and some punctuation, as with what became conventional typewriters, this device could produce pixelated versions of any desired symbol. For example, Figure 5.9 presents some Devanagari letters, ligatures and words using some made up codes within the same general approach. This is an example of direct applicability of a writing machine to a Brahmic script, most of which have characteristics making them generally ill-suited to what became conventional typewriters, as discussed in Chapter 3.

Earlier we discussed the use of the Smith Premier for Thai, which, like Devanagari, is a Brahmic abugida script. Even the double keyboard typewriter, however, would have had difficulty handling aspects of certain abugidas, such as complex compound and conjunct ligatures that occur when certain symbols are combined. Figure 5.9 shows the letters प ⟨pa⟩ and त ⟨ta⟩, the ligature त्र ⟨tra⟩ and the Hindi word पत्र ⟨patr⟩, which translates to 'letter' (as in correspondence). Note that the त्र ligature represents the sound /tra/ (or just /tr/ with schwa deletion). When typing on a computer, this can be achieved when the letter त (⟨ta⟩) is combined with the *virama* to remove the vowel, resulting in त् (⟨t⟩) which itself is then combined with the letter र (⟨ra⟩), resulting in the irregular ligature त्र.

In Chapter 3 we presented details on Devanagari typewriters, which, as with other Brahmic script, required decomposing many characters into separate sub-characters, so that, for a significant number of characters, multiple keys would be used to produce a single full character. These keys would be selected in sequence, one at a time, and their composition was enabled by not advancing the carriage for some subset of sub-characters. Thus the degree of character decomposition was much more limited than with the pixel-level Raphigraphe; and, while multiple keys were required per character, they were not selected simultaneously (due to mechanical constraints on type-bar collisions). The characters in Figure 5.9 would likely be typed much more quickly and easily with the Devanagari typewriter than the Raphigraphe.

However, the point remains that the Raphigraphe writing machine from 1843 is straightforwardly extensible to representing the Devanagari writing system and likely any other. In use it may not be fast and the resulting pixelated text might not look great, but it would be functional. The key reason that the Raphigraphe never caught on is the complexity of producing a single letter, which is directly related to the flexibility it has to produce arbitrary symbols as demonstrated above. Other mechanical typing devices were designed in the 19th century that occupy a different operating point on the efficiency/flexibility continuum, while making different choices than the one-key-one-letter paradigm that became dominant. In particular, the Livermore (1863) mechanical typographer device had only 6 keys, and letters were specified by pressing multiple keys simultaneously—same as the Raphigraphe—but only once per letter, rather than for multiple columns. To accomplish this, four of the keys were associated with straight lines (left, top, right, bottom) and two keys were associated with upward (∧) and downward (∨) angles. Figure 5.10 shows a schematic of how each of these lines were assigned to the 6 keys—three for the left hand and three for the right— which form the basic components of each letter. Figure 5.11 presents what the alphabet looks like using the device, according to the Livermore (1863) patent.

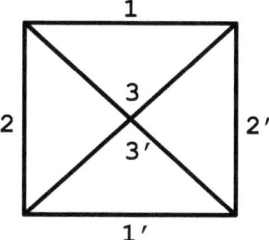

Fig. 5.10 Lines associated with each of 6 keys in Livermore (1863) mechanical typographer.

It is clear from the presented alphabet that liberties were taken with the conventional form of the letters in the Latin script, e.g., the letters A, E and F lack the mid-line that is typical in their uppercase form, and the letters S and T do not particularly resemble their conventional forms. We note that in Livermore's figure, the reference letters beside each of the forms produced by his device are a mixture of uppercase and lowercase forms, whichever most closely resembles the device version. Inspecting the schematic, we can see that the way to type the letter h with the device would be to simultaneously press keys 2 and 3', i.e., the second key with one hand and the third key with the other hand.

Comparing this approach with the Raphigraphe, we can make a few observations. First, the device would be much easier and faster to use. However, the resulting letters are worse approximations of conventional printed characters. This is because the Raphigraphe had the flexibility to produce forms at the pixel level, while Livermore's device relies on fixed line segments. To quantify these differences, the total possible combinations of pixels in a single column of the Raphigraph is $2^{10} = 1024$, which, since we combine multiple (say 5) columns for each letter, results in a set of 2^{50} (1.1 quadrillion) possible forms. In contrast, the Livermore device allows for $2^6 = 64$ possible forms. So no easy extension to Devanagari.

In the next chapter, we will examine the question of speed of text entry, and one technology that has successfully exploited the use of multiple simultaneous key presses per character is the stenographic keyboard, which is specifically intended for the rapid entry required for live spoken language transcription. The conventional typewriter keyboard, however, is the result of several design choices—one-key-per-letter, three rows of letters that are selected with a single direct selection, with letters ordered in a conventional (and now familiar) way. With the advent of computers, and machine-readable text, input methods that make use of as few as just a single key (similar to the methods presented at the beginning of the chapter) become possible, making writing accessible to those who would not otherwise be able to write. Similarly, possibilities for general speed improvements exist, and will be the

Fig. 5.11 Alphabet in Livermore (1863) 6-key mechanical typographer. Work is in the public domain in its country of origin and other countries and areas where the copyright term is the author's life plus 70 years or fewer.

topic of the next chapter. Before getting to that, however, we review the technology and how it can improve accessibility.

5.4 Machine readable text entry

The technologies that have been discussed in this chapter up to now have been manual or mechanical in nature, meaning that the correspondence between the required physical gesture and the resulting symbol are direct. The transition to machine readable text made possible with computers allows for less direct methods of text entry, which have become prevalent in many of the most common communication modes today. Most of us take such text entry assistance for granted, but it has opened up possibilities for text-based communication that did not exist before, allowing individuals to independently communicate who would not otherwise be able to do so. Before turning to such life changing options, we shall first examine what are now ubiquitous quotidian text entry technologies.

Note that in many cases we will be describing text entry methods that rely on some form of artificial intelligence (AI), however basic, to even allow the method to function as a text entry system. Additionally, AI can be used to speed up text entry, a key topic of the next chapter. In this chapter we will discuss only those AI methods related specifically to input gestures, and defer discussion of other AI uses (even sometimes within the same systems discussed here) to Chapter 6.

5.4.1 Representation versus rendering

With a mechanical typewriter, such as the famous Remington shown in Figure 5.2, pressing the 'w' key on the keyboard directly yielded a 'w' form on the page via the type bar striking the paper. A physical keyboard attached to a computer does something quite different when the 'w' key is pressed: it stores the letter in an internal intermediate representation for subsequent use. What that internal representation is does not really matter as long as the resulting text can be presented to the message recipient. In fact, there are many ways in which a computer might represent letters from the Latin script (including 'w') internally, and quite a few that were actually adopted for real systems. Abstracting away from that detail (to which we will return shortly), this kind of internal representation allows us to distinguish between the representation of the text and the rendering of the text, which we will illustrate next through some simple examples.

Perhaps we pressed the 'w' key to enter the last letter of the English word 'flow'. Each of the four letters in the word are represented as a code, which are accessed when the word is presented to the recipient, e.g., by printing onto paper or via an electronic display. The word can be rendered during presentation in many different ways. Text rendering is the process by which the underlying letter form is presented visually, which may differ in key ways (size, font, etc.), even with the same underlying representation of the letter. Old typewriters allocated a fixed width to each letter, and we can choose to use such an approach with machine readable text, too: `flow`. We could choose to emphasize the word by italicizing, bolding or increasing the size: *flow* **flow** flow!!! Because the text is stored in an internal representation, we could also apply different kinds of processing to the word prior to presentation, such as converting it to all capital letters (FLOW) or even translating it to French (couler). Hence while seemingly a minor difference—storing the intended letter in some intermediate representation versus directly producing it—this separation between the representation and rendering—opens up many possibilities.

The ASCII encoding system was included in the Unicode standard, hence the letter 'w' is encoded as `01110111`. In many scripts, some glyphs are represented in Unicode by a single code point, i.e., the stored code for that glyph identifies the glyph as a whole. Others, however, may involve the combination of multiple code points, and the rendering engine has to know how to combine them together. For example, certain diacritics are combined with a base letter in a single code point, while other diacritics are code points in the Unicode system in their own rights, and must be independently specified in order to combine with a base letter. Sometimes there are multiple ways to represent the same glyph—e.g., with a single Unicode code point as well as with the base letter plus a separately encoded diacritic, both resulting in visually identical glyphs. Various normalization forms, such as Unicode's

Normalization Form C (NFC), were created so that strings represented by alternative valid encodings are converted to the same canonical encoding.

For example, the character 'ǭ' (U+01ED) is part of the Latin Extended-B block of Unicode, and combines the lowercase Latin letter 'o' with a macron diacritic above and an ogonek diacritic below. It is used in the Osage alphabet, a Latin script writing system for the Osage language.[12] Both of the diacritics used in this single code point have their own individual code points in the Combining Diacritical Marks block of Unicode—Combining Macron (U+0304) and Combining Ogonek (U+0328). As a result, this particular character can also be rendered from a combination of the base lowercase 'o' with these two combining characters, in either order. Unicode Normalization Form D (NFD) decomposes a character into its canonical sequence of base characters, in this case Latin Small Letter O (U+006F), Combining Ogonek (U+0328), Combining Macron (U+0304). Unicode Normalization Form C (NFC) first applies NFD then composes the base characters into canonical composite code points, in this case the character 'ǭ' (U+01ED). By applying NFC to text, any possible representation of 'ǭ' will be converted to that particular code point. This can be important when deciding, for example, whether words that have different byte encodings are in fact equivalent.

If the goal is to provide a canonical representation for all strings that are visually identical, then additional normalization beyond NFC may be required. For example, if an additional Combining Ogonek diacritic (U+0328) follows the character 'ǭ' (U+01ED), which already has an ogonek diacritic below, then visually nothing changes in the rendering of the text. However, NFC normalization does not remove the extra diacritic in the text encoding, so that two distinct text strings (<U+01ED> and <U+01ED, U+0328>) produce the same rendered text. Conversion of text containing such visually inert material to a more canonical encoding is one component of what is termed *visual normalization* by Johny et al. (2021), which documents software utilities that, among other things, perform such conversions in a range of Brahmic scripts, where such issues are particularly acute.

All methods for machine-readable text entry map user input gestures to an internal representation before doing anything with that representation, such as displaying, printing, sending or speaking it. This creates a tremendous flexibility in designing text entry methods relative to the mechanical methods outlined earlier, which had a direct correspondence between inputs and outputs. As long as a system can correctly recognize that a particular sequence of input gestures corresponds to a particular string of letters, it does not matter whether the user directly selects those letters on a keyboard or uses some other, less direct method to specify them. Sometimes less direct input methods have advantages over direct methods.

---

[12] In addition to this Osage alphabet, a distinct Osage script was developed in 2006 and added as a separate block to Unicode in 2016.

5.4.2 Conversion to machine-readable text

We discussed earlier various recognition technologies—speech recognition and handwriting recognition—that convert user input in the specified modality to machine-readable text. In practice, speech recognition systems take dictation and handwriting recognition systems play the role of typist, much as humans performed such tasks over the past centuries and millennia, producing documents directly from author speech or handwriting. We can add printed or typed text to this list of modalities that can be converted to machine readable text, in this case via what is known as optical character recognition. All of these automatic recognition technologies have reached impressive recognition accuracies in many common use scenarios, though (as with humans) some errors are to be expected, hence document review and editing is typically recommended even in such common scenarios. In scenarios where the input is less typical—dysarthric speech, difficult to read handwriting, or blurry printed documents—more errors (hence more editing) may be expected.

One way to think about recognition technologies is in terms of system coverage, i.e., the range of inputs to the recognition system for which relatively accurate (sufficient to be useful) conversion to machine-readable text is possible. Coverage may be assessed over languages supported: speech recognition systems are commonly quite capable of accurately recognizing many commonly spoken languages (English, Chinese, Spanish, Arabic, etc.) but no system recognizes all, or even most languages. In fairness, this is also true of humans taking dictation—speaking a language is a prerequisite for accurate recognition and transcription, and nobody speaks every language. This is not only the case for speech—recognizing handwritten or printed Arabic, for example, will be beyond the capabilities of humans or machines without the requisite training. Coverage can also be assessed in terms of scripts, e.g., Punjabi handwriting may be recognized in the Gurmukhi script but not in the Shahmukhi script; or in terms of fonts or typefaces—a system's optical character recognition of historical English may be accurate in a Roman type but not in a Gothic type (or Blackletter). Noise conditions can also impact coverage, for many sources of noise that might obfuscate the input. In all of these cases, coverage issues exist for people as well as for machines. And in most cases, machine coverage is improving over time to include more and more languages, scripts, fonts and conditions.

It has always been the case that recognition technologies restrict coverage in key ways to yield useful systems, and growth in coverage occurs when the technology reaches high enough accuracy under existing coverage restrictions that effort can be focused on new challenges. We do not have the space to cover the history of these recognition technologies (see, e.g., Jurafsky & Martin, 2008, and references therein), however within the sphere of text entry, we can discuss a couple of cases where restrictions on input were imposed as the means of achieving useful narrow-coverage systems.

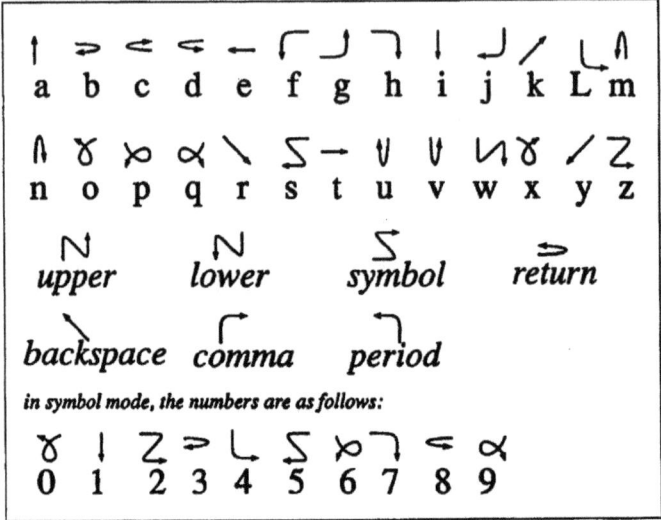

Fig. 5.12 Recognized strokes in the Unistrokes system, graphic from Goldberg (1997). Source: patent.

One such system was already discussed earlier in the chapter—the NASA speech recognition system for use on the International Space Station. This system did not provide accurate recognition over the entire English language, rather had a restricted vocabulary that it could accurately recognize. This fit the specific needs of the system—not general dictation but rather some kind of command and control interface. Similar vocabulary restrictions are typically also in effect for interactive voice response (IVR) systems that provide automatic assistance over the phone. However, general spontaneous speech recognition has improved dramatically in recent years so that explicit vocabulary restrictions are not required for many languages in typical conditions.

Input restrictions have also been used for handwriting recognition systems. In this case, restrictions have been placed on the shapes of letters that would be recognized. Perhaps the best known such restricted input system was Graffiti, which was used on Palm mobile devices beginning in the mid-1990s. The system was similar to the Unistrokes approach from Xerox,[13] which we present in Figure 5.12. For each target glyph (alphabetic, digits, comma, period) or operation (case shift, return, etc.) there is an associated stroke that is drawn with a stylus. The strokes are pulled from a relatively small vocabulary which was designed to be easy to input without looking at the device (similar to touch typing in this respect) and also easy for the system to recognize, i.e., distinguish the strokes from each other. All of the strokes are input without lifting the stylus from the surface of the device. While some of

---

[13] Indeed, Palm eventually settled a patent infringement lawsuit brought by Xerox over the similarities.

the strokes are visually related to the glyphs being input—such as 'i', 'L', 's' and 'z'—most are sufficiently distinct from the target glyph that users had to learn the system in order to input text effectively.

The Unistrokes bear some resemblance in certain cases to Livermore's (1863) alphabet previously shown in Figure 5.11, which also made use of a relatively small number of primitive lines to compose letters. They more closely resemble Moon's alphabet for the blind, a graphic of which from Moon (1873) is presented in Figure 5.13. Moon's alphabet, an alternative to Braille's more popular system of embossed dots, was initially printed (not typed—typing devices for the alphabet came much later), and was designed to be easy for readers to distinguish between the letters and recognize them even without much training—motivations related to those of Unistrokes.

Continued rapid improvements in AI systems means that the coverage of such systems—in terms of languages, scripts, and conditions for which useful conversion to machine-readable text is possible—is constantly expanding, hence text entry options for those using such technologies will only increase. Challenging conditions, however, will always exist—much as they do for human's ability to process language—and these will require specialized solutions, such as the discussions of dysarthric speech recognition in Section 5.2.1.

5.4.3 Ambiguous and virtual keyboards

Conventional physical keyboards that are attached to a desktop computer or built into a laptop retain a mechanical component and commonly borrow much of their surface design from the conventional QWERTY typewriter keyboard. In our discussion of mechanical typewriters, we highlighted the assumption of one key per letter in the conventional typewriter design, and presented a few (largely forgotten) alternative typewriter designs that departed from this assumption by requiring multiple keys to be pressed to produce a single letter. Another way to depart from this assumption is to assign multiple letters to each key, which was a common way to provide text entry on early mobile telephones with keypads that had only 12 keys total. For example, a schematic representing a common 12-key layout—four rows of three keys per row—is presented in Figure 5.14. In this layout, three or four Latin script letters are assigned to numeric keys 2-9 in alphabetic order. This keypad served dual purpose on such phones—allowing phone numbers to be input directly, and letters to be specified ambiguously, making this an instance of what is known as an *ambiguous* keyboard.

Dealing with the ambiguity of having multiple possible intended letters assigned to single keys is something that machine-readable text entry greatly facilitates, since, as discussed earlier, these systems can operate on intermediate representations to determine the intended text. In this case, if a number on a 12-key keypad is chosen, additional input can be elicited from the user

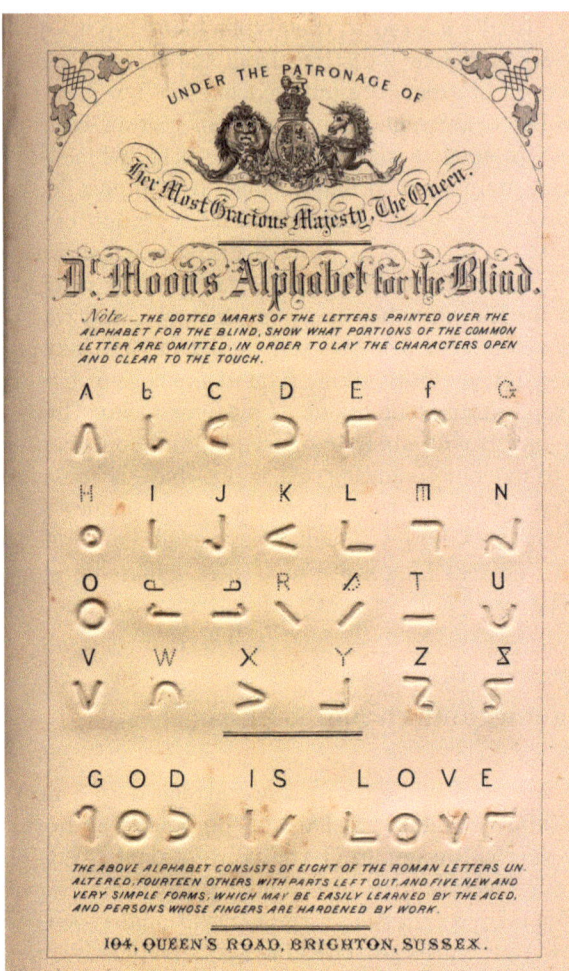

Fig. 5.13 Moon's embossed alphabet for blind readers, originally proposed in 1845. Work is in the public domain in its country of origin and other countries and areas where the copyright term is the author's life plus 70 years or fewer. Source: Moon (1873).

to disambiguate between the possible letters. MacKenzie & Soukoreff (2002) present several methods that were used commercially for such disambiguation. One common method, known as 'multitap', required a different number of keypresses (or taps) to select the different letters: one tap for the first listed letter, two taps for the second, etc. A brief timeout can be used to indicate that the selection is complete, which is important because without it the input gesture can be ambiguous. For example, to type the word 'back' with the multitap method using the keyboard in Figure 5.14 requires the following input: 2, 2, 2, 2, 2, 2, 5, 5 (two 2's for 'b', one 2 for 'a', three 2's for 'c' and two 5's for 'k'). This is the same input required for 'cck', an abbreviation that

1	2 abc	3 def
4 ghi	5 jkl	6 mno
7 pqrs	8 tuv	9 wxyz
*	0	#

Fig. 5.14 Schematic representing a 12 key layout common on early mobile phones. Source: own work.

can be used for, among other things, the peptide hormone cholecystokinin. To distinguish between these (and other) possible intended letter strings, a short pause can be inserted after the multitap entry for each letter if needed.

A second method to disambiguate requires exactly two keypresses per letter, hence is referred to as the two-key method. First press the key with the letter on it, then press the number corresponding to the position of the letter on the first pressed key. Thus 'back' would be input: 2, 2, 2, 1, 2, 3, 5, 2, i.e., the second letter on the 2 key, the first letter on the 2 key, the third letter on the 2 key, and the second letter on the 5 key. This sort of numeric selection method is often used when a list of options is presented to the user, such as in certain pinyin input methods for Chinese.

A final method mentioned in MacKenzie & Soukoreff (2002) is to add a modicum of AI to the system by ranking the possible input words by probability and only require additional effort if the desired word is not the most probable from among the alternatives. The most well-known such approach was called T9,[14] and it relied upon a lexicon of words in English with their frequencies. The keys are pressed just once for each intended letter (e.g., 2225 for 'back') and the most likely corresponding word in the database is presented. If that is the desired word, then the space key (typically 0) could be pressed; otherwise another key (often *) was pressed to display the next most likely word, until the desired word is presented, at which point space can be selected. For very frequent words, this will be a very efficient way to enter the text. Further, new words input by the user were typically added

---

[14] Note that the initial T9 publication (Kushler, 1998) presented the approach as "a new technique for augmentative and alternative communication text input" rather than for general mobile text entry, though it is now more associated with the latter.

to the lexicon, so that some degree of personalization was provided by the system. Issues around the use of predictive models in text entry—including personalization of such systems—is also addressed in the next chapter.

Virtual keyboards do not consist of physical keys, but are rather rendered for presentation, typically on a computer or mobile device screen. This style of keyboard is now very common on mobile phones and tablets with touch screens, and most current virtual keyboards are constrained by limited screen size and make use of predictive models to improve speed, both topics that are key topics in the next chapter. Yet there are some aspects of the technology that are worth touching on in this chapter, given the importance of input gesture for such keyboards.

Among the earliest virtual keyboards were those used as an alternative input modality to the previously discussed Graffiti/Unistrokes type stylus strokes on personal digital devices. A small QWERTY keyboard was represented virtually on the screen of the device, and the same stylus used to produce Graffiti strokes could be used to select letters on the virtual keyboard. This sort of direct selection by touching with a stylus is akin to so-called 'hunt-and-peck' methods on typewriters or physical keyboards, and requires attending to the virtual keyboard for accurate selection, something that stroke-based input was intended to avoid. These days selection with a stylus has been largely replaced on mobile devices with selection via thumbs and other fingers, something requiring extra intelligence within the virtual keyboard to manage the corresponding lack of touch precision.

One method combining the naturalness of simple stylus strokes with a fixed position virtual keyboard like QWERTY was what became known as gesture input (Kristensson & Zhai, 2004). The idea was to create a single stroke that started at the first letter of a word, traced through every letter in the word, and finished at the last letter of the word. For example, to type the word *this* on a QWERTY keyboard using this method, the user moves the stylus from the letter 't' on the keyboard, through 'h' en route to 'i' and then back to 's' without lifting the stylus. Using technology similar to the stroke recognition systems used for Graffiti or Unistrokes, whole words could be recovered from such gestures without requiring the selection of each letter individually in the sequence. As with direct letter selection, this input method has persisted to the current day—such input can be found on many virtual keyboards for mobile phones—albeit typically using fingers rather than a stylus. Recognizing such gestures on mobile devices given the imprecision of finger touches on very small keyboards requires system intelligence, which we will return to in the next chapter.

5.4.4 Editing controls

Another major difference between machine-readable text entry and mechanical typing or handwriting is the inclusion of editing controls as part of user input. Some editing controls have a history in mechanical methods, but the interface flexibility when revision is allowed makes such controls indispensable. On a typewriter, a backspace key—which eventually became common—allowed the carriage to move one position backwards without adding to what was typed on the page. This is literally the reverse of typewriter space bar functionality, which simply advances the carriage by one position—hence the term backspace. On typewriters, this movement allows things like adding accents above letters or other additions to what had already been typed. This purely navigational functionality is duplicated on modern physical computer keyboards with navigation arrows that allow the cursor to be moved around a message/document without altering the text. The term 'backspace' has now become typically synonymous with 'delete' in that it not only moves the cursor back one position but erases whatever text was in that position. While some advanced typewriters (near the time of their ultimate obsolescence) did provide some editing functionality, the degree of editing control is significantly higher with machine-readable text entry, as is its importance within interfaces.

Performing editing operations on already entered text is part of the set of user input gestures that must be accounted for within a text entry system. This was shown (but went unremarked) earlier in Figure 5.12, which displays the Unistrokes system. In addition to the letters, digits and punctuation, there were gestures for lower/upper casing, return and backspace. Here backspace has the semantics of deletion, i.e., it removes the last typed symbol, which, as we mentioned, is how it is typically used in contemporary virtual keyboards as well.

Inclusion of even a relatively basic and seemingly straightforward editing operation such as backspace/delete increases some of the complexity of the interface. One ramification of this in particular is that there is no longer a one-to-one relationship between the user input gestures and the resulting text: different sequences of input gestures can lead to the same resulting text. If the text string produced is the word *cat*, then this probably resulted from the user selecting the letters 'c', 'a' and 't', but it might have resulted from the user selecting the letters 'c', 'a' and 'r', followed by backspace/delete and 't'. Or possibly other deletions were made in the string—this cannot be determined simply by examining the resulting string of letters.

This raises some relatively subtle points that must be accounted for in the design of the text entry system. First, regarding the semantics of backspace/delete, is this operating only on the text that has been produced, or on the sequence of input gestures? The latter semantics is more like an 'undo' operation, so that two consecutive backspace/delete inputs would cancel each other out, making it impossible to delete more than one letter from the end of the entered text. Hence for a simple text entry system, backspace/delete can only operate on the output string and no 'undo' functionality is provided. Such design choices can become more problematic when predictive models are used within a text entry interface, where undoing earlier selections from provided predictions can be an important option.

While on the topic of predictive models for text entry, one might expect that such models could be used to predict when a user is likely to delete or modify some already entered text. However, the fact that editing operations are not observed directly in text means that the raw data used to train predictive models is missing information about how the text was actually produced. As with the *cat* example above, many common errors and/or misspellings are corrected on the spot and there is no evidence in the resulting text that this was the case. The fact that, in English, 'u' is a likely letter to follow after 'inq' has been typed (and 'z' is not) is something that can be observed in a large corpus of text in that language. How likely a backspace/delete is after that string of letters is harder to estimate since that is not explicitly included in the large text corpus. This is another topic we return to when we look at the use of predictive models in the next chapter.

More complex editing gestures—such as selecting some text earlier in a document and deleting or modifying it—is harder to accomplish in some interfaces than it is in others. These editing gestures (direct pointing into the text, via a mouse or with the finger, specifying a span of text at that position to modify, and, finally, editing it) typically require relatively fine motor skills and can be far more time consuming than non-editing operations. How to permit such document editing when the user cannot produce the complex gestures typically required to do so also requires careful design, which leads to our next topic.

5.5 Augmentative and Alternative Communication

We began this chapter by presenting a scene from a television drama in which someone with a severe motor disability communicated a message to a caregiver by selecting presented options via a small bell. The field of Augmentative and Alternative Communication (AAC) is focused on providing communication methods for individuals who, for whatever reason, cannot use speech unaided for communication needs. Practitioners in the field (often speech-language pathologists) provide strategies and technologies to individuals facing such challenges, due to disabilities (permanent or temporary) of whatever kind. These strategies may or may not involve computer technology, which is a key distinction in the field. Many methods that do not involve computers—so-called low-tech approaches, such as the one illustrated at the start of the chapter—are effective for immediate communication of needs, to caregivers or others. These typically involve direct interaction with the communication partner during message construction, as is demonstrated in the example scene. So-called high-tech methods involving computers can allow individuals to construct messages independently, which may then be presented to the communication partner via text-to-speech synthesis or just as text. However, typical text-entry methods (typing, etc.) may also be difficult due to the same speech-impacting disabilities. For example, cerebral palsy can impact both speech intelligibility and the motor control required for direct selection of the keys on a keyboard. Specially designed methods for text entry can allow individuals with such disabilities to effectively control their own message creation, although speed of input may be slow.

We conclude this chapter on gesture complexity with a brief presentation of the kinds of user input gestures used for text entry in such scenarios. Many of the most important technologies in the field ultimately rely heavily on AI to help speed up text entry as much as possible, hence we will revisit this domain as a key consideration in the next chapter.

5.5.1 Speech generating devices

A speech-generating device is a computer interface allowing for both text entry and production of synthesized speech from the entered text. This is a very general definition that now includes most laptops and tablets, although historically the term has been used for specialized devices that provide text entry interfaces that can be tailored to the needs of individuals with severe speech and language impairments. Much of this tailored functionality can now be provided by apps on general purpose devices, though the task of setting up the communication system to fit each individual's needs and capabilities requires the availability of extensive configurability of devices and apps.

The need for a device of this sort for communication arises in many very distinct situations, for individuals often with relatively limited capabilities, so flexibility of interface is at a premium. Unlike other text entry scenarios discussed in this chapter, potential lack of user literacy is something that must be accounted for in such interfaces. Pre-literate children or adults who are functionally illiterate may have the need for a speech generating device to communicate, for all of the same reasons that any of us need to communicate: to convey information, to request help, or even to tell a story (Black et al., 2010). Literate adults may require such a device due to either congenital or acquired disabilities, so that some individuals may have had the time (and inclination) to learn new methods for text entry, while others may want to rely on familiar and/or easy to use interfaces even if those will be slower than more complex alternative methods that require training. Sometimes there may be important or frequent full utterances that users would like to be able to select quickly, such as "Hi, how are you?" or "I need some help." Other times an individual may want to take the time to write something from scratch—an email, an article or even a book. The ability to quickly select words, phrases or even full utterances must be balanced with the flexibility to construct new utterances letter-by-letter if necessary.

Beyond literacy and communication intent, individuals differ in their physical capabilities, so that the gestures for interacting with the system may need to be tailored to what the specific individual is able to reliably produce. If the easiest movement for an individual to consistently perform is raising an eyebrow, then eyebrow raising should be available as a means for interacting with the system. At the beginning of the chapter, we presented a scenario with an individual who could ring a bell with their index finger—a similar gesture should be available for text entry in a speech generating device. Speech-language pathologists work with individuals to determine the easiest, most reliable gestures that the individual can provide to the system, then make use of sensors and switches that can capture those gestures to interact with the system. Eye or head movement, muscle twitch, direction of gaze, and hand or foot gesturing are all commonly used to drive text entry in such systems (Beukelman & Mirenda, 1998). Even electrical activity in the brain can be detected on the scalp and used to drive text entry without any voluntary movement required (e.g., Oken et al., 2014). The required switches and sensors are typically connected to the device as peripherals, much as a keyboard attaches to a computer. Then the text entry software must be able to make use of this input to produce the utterance.

This very general description of how text entry operates masks the extreme degree of personalization that is required in AAC, which is accomplished by professional speech-language pathologists, who are experts in deploying technology in real-world settings. In the next section, we continue the general characterization by framing text entry as (1) presentation of items to be selected; and (2) methods of selection.

5.5.2 Text entry as selection

Figure 5.1 at the beginning of the chapter presents the AEIOU letter board that was used in the Breaking Bad scene to elicit the utterance "need DEA" from Hector Salamanca. This 6-by-6 grid shows letters and digits for spelling out words. In general terms, the items presented in the grid are part of what is called the 'selection set' and in this case it is fixed, meaning that it does not change as letters are selected. What to include in the selection set is also something that the speech-language pathologist will work with the individual to determine. For example, a selection set could consist of icons representing topics. A set of possible topics would typically be a *dynamic* selection set, meaning that once a topic is selected, a different (topically relevant) selection set would be presented to the individual. For example, if the topic "Food" is selected, a new selection set would be presented, potentially including items such as "I'm hungry", "Breakfast", "Lunch", "Dinner", "Dessert", "Fruit", "Pizza" or other food-related utterances or sub-topics that were chosen for the "Food" selection set. For the topic "Music", a distinct selection set, ideally also geared to the individual's taste, would be presented.

Common forms of hybrid fixed/dynamic selection sets are virtual keyboards with a standard fixed keyboard portion that displays the letters in the QWERTY layout and a small region (often above the letter keys) allocated to possible next words or word completions. Predicted word completions change with every letter input (hence this region is dynamic) but the QWERTY part of the virtual keyboard does not change—unless a shift key is selected, causing the fixed portion of the keyboard to display an alternative fixed selection set.

Given a selection set, text entry proceeds by selecting items from the set. Selection can occur directly by pointing at the intended item, and methods for pointing can involve making contact with the item (touching or pushing down) or not. Contactless methods include manual pointing, sustained eye gaze at the target item (which requires gaze tracking) or the use of a laser or light source directed at the target. Most of these methods require some way of distinguishing incidental pointing or gazing versus instances intended to select the item. This is typically distinguished based on the duration of dwelling on an item, which can be configured to the abilities of the individual. For example, if eye gaze is used to select items from the selection set, the individual may scan the set to find the target item, then fix their gaze on that item until the requisite dwell time is reached, at which point the item is selected.

Indirect selection methods do not require such targeted selection gestures. The scenario presented at the beginning of the chapter was an instance of (low-tech) indirect selection. The individual was apparently unable to make consistent pointing gestures to items on the presented grid, so the communication partner instead asked yes/no questions to determine each target item. As we stated earlier, the answers to the yes/no questions are effectively a code to specify the item, and indirect selection methods will differ in how such codes are established and indicated.

In computer interfaces, the most common form of indirect selection is known as scanning. Scanning involves sequentially highlighting subsets of items for selection in a way that eventually allows the selection of a single item from the set. Highlighting a subset of items can be interpreted as a question: is your target item in this subset? The most straightforward scanning method is linear scanning, which involves highlighting each item by itself, thus requiring just one positive selection gesture per item. For example, given the letter grid from the start of the chapter, each letter could be highlighted in alphabetic order until the target letter is highlighted, at which point a 'yes' answer can be provided by the user. This would be a pretty fast method if the target letter was 'a', 'b' or 'c', but not so fast for 'x', 'y' or 'z'. Another linear scanning approach would be to highlight letters in frequency order, so that common letters are highlighted (hence selected) earlier.[15] In contrast to linear scanning, group/item scanning methods first highlight multi-item subsets, then when a subset is selected, highlight individual items in the subset. One common type of group/item scanning is called row/column scanning, which is what the caregiver used in the cited Breaking Bad episode, whereby rows in the grid are presented for selection, and once a row has been selected, letters from the row are presented. Roark et al. (2013) introduced methods for choosing non-contiguous subsets of letters to highlight in fixed-grid group/item scanning that created optimally short codes from a predictive model. Their method highlighted subsets in more dynamic patterns than the contiguous rows used in row-column scanning, which required more attention and effort from individuals than row/column or linear scanning, but did lead to text entry speedups in the experiments presented in the paper.[16]

For any scanning methods used for indirect selection, parameters that control the scanning process must be established. First, with a single yes/no switch, how does scanning proceed and how is selection made? Automatic scanning dwells at an item (e.g., a row in row/column scanning) for some duration before moving to the next one; and selection is made by activating the switch. This is the method that was used by the caregiver in Breaking Bad. Step scanning inverts the semantics of switch activation from automatic scanning: it advances to the next option when the switch is activated and selects an item if the switch is not activated after dwelling for some duration. Directed scanning begins when the individual activates the switch (e.g., pushes down on a button) and selection is made when they release the switch. In each of these cases, the time that the system dwells on an item to allow for selection is a key parameter of the system, and depends on the capabilities and desired speed of the individual. Some users learn the patterns associated with particular letters so well that they can set the dwell time to be very short, much shorter than would be possible if one were reacting to the highlighting rather than anticipating it. Long-term users of such technologies, such as Stephen Hawking, may eventually use dwell times that are scarcely perceptible to novice users or those who do not use the technology.

We note that, as with other styles of virtual keyboards, AAC systems must provide editing controls in addition to alpha/numeric inputs, something that our example 6-by-6 grid earlier in the chapter omitted. If the individual makes a mistake in selection, they will need some way to repair this error, and individuals making use of these technologies will in general have the same array of editing needs as anybody else.

For more details about these and related topics in AAC, we refer the reader to Beukelman & Mirenda (1998). In the next chapter we will return to discuss AAC while presenting methods making use of predictive models, which are

---

[15] This is how the partner-assisted scanning was performed by Jean-Dominique Bauby and his communication partner, as discussed in Chapter 1.

[16] See Section 6.4.3 for further discussion of this method.

of particular usefulness in AAC due to the slow speed of these text entry methods for many people.

Chapter 6
Writing under space and time pressure

6.1 Prelude

Skill at the physical act of writing—originally handwriting, later typing or "word processing"—was until relatively recently the ticket to a steady job. While such jobs have not completely disappeared, copyist professions such as that described in Herman Melville's 1853 short story "Bartleby, the Scrivener: A Story of Wall Street" are no longer commonplace. The story concerns a law office employing legal copyists, who were paid to produce copies of legal documents "at the usual rate of four cents a folio (one hundred words)". The narrator of the story describes one employee as somebody who "wrote a neat, swift hand", and provides this description of a typical piece of work:

> A few days after this, Bartleby concluded four lengthy documents, being quadru-plicates of a week's testimony taken before me in my High Court of Chancery. It became necessary to examine them. It was an important suit, and great accuracy was imperative. Having all things arranged I called [the clerks] from the next room, meaning to place the four copies in the hands of my four clerks, while I should read from the original.

Being paid by the word for neat and accurate transcription provides a strong motivation to write (or type) as fast as one can while maintaining the required quality. Producing four 'lengthy documents' in the time it would otherwise take to produce, say, two would have real monetary consequences. Those taking dictation, thus needing to transcribe at the speed of speech, would also be motivated to pursue methods that allow them to write/type fast enough to keep up. These are skills that require training and for which there was sufficient demand to make the training worthwhile.

In addition to speed in writing, expertise in the precision of the writing gesture also led to (somewhat less commonplace) professional opportunities, mostly related to the novelty of very small writing. The common idiom "in a nutshell" derives from a claim by Cicero (according to Pliny the Elder) of encountering a version of Homer's Iliad written in such a compact way

B. Roark et al., *Tools of the Scribe*,
https://doi.org/10.1007/978-3-032-00831-2_6

that the parchment upon which it was written was housed in the shell of a nut. While there are practical reasons for writing compactly—including using less paper, which was cited in the prior chapter as an early rationale for typewriting—writing with the degree of compactness presumably required for Cicero's Iliad example (unless the nut was a coconut) has little practical value, but we still speak of it a couple thousand years later.

Novelty small writing persisted into modern times. The Reading Times newspaper of Reading, PA, printed an event notice on Oct. 15, 1934, entitled "Freaks of World Fair Open Engagement Today". Robert Ripley—originator of the "Believe it or Not" syndicated comic highlighting bizarre and unusual 'facts'—sponsored an exhibit at the Chicago World Fair in 1933 called "Ripley's Odditorium". The exhibit was reported to be so disturbing that beds were made available on-site for anyone who fainted. After the World Fair, the spectacle was taken on the road, and the article announcing the opening in Reading states:

> Among the living oddities who will be seen here will be.... Harry Overdurff, the man who is slowly turning to stone; Elma Von Lynd, the three-legged girl; J.B. Shuster, the only man in the world that can pickup 20 baseballs in one hand from a flat surface;... Blystone, known as (Bly) the rice writer, the only man in the world to write 2786 words on a grain of rice; Mayfield, the human torch …

One may be left with questions from this teaser,[1] however, while seeing Bly the rice writer may not have induced fainting, witnessing a demonstration of the skill required to perform such a transcription[2] was apparently enough of a draw to get Blystone a position on the tour.

Microscopic writing does have its uses, such as in steganography, i.e., when the text is intended to avoid detection. However, such uses have long had technological solutions that do not require the motor skills involved in manual microscopic writing. For example, Knight (1877) presents schematics of an 1852 invention—the Micropantograph—which mechanically reduces the size of handwritten text by nearly four orders of magnitude. Using such a device, it was claimed in 1862, allowed 223 letters to be etched onto 1/356000 of a square inch of glass, by a device operator making more conventional scale handwriting gestures. Other technologies, such as microfiche, allow for document duplication at minute scales, achieving much the same purpose without the specialized writing skills. As a result, the kind of skills that Blystone exhibited in the Ripley spectacle are mostly to be thought of as special skills of dexterity, along the lines of what Thea Alba was known for demonstrating (e.g., simultaneously writing multiple distinct messages with different hands, feet or fingers—see, for example, Jay, 1986). So, unlike fast and neat writing,

---

[1] Was an oddity's impressiveness inversely proportional to how many words it took to describe it, or does it just seem that way? Was that a grain of Basmati or Arborio rice? How many words were written on a grain of rice by the runner up?

[2] The newspaper was mistaken: Blystone's actual feat involved that many letters not words.

this skill was not a typical route to a profession, and apparently well suited to Ripley's Odditorium.

There was, however, one class of compact writing methods that was in high demand—methods that, due to an increased economy of gesture, contributed to transcription speed. Returning to a key part of the Melville story quote above, unremarked upon until now: "...four lengthy documents, being quadruplicates of a week's testimony taken before me in my High Court of Chancery." Bartleby's employer is the narrator of the story and presumably transcribed the testimony as it was being spoken in court. To transcribe speech as it was being spoken typically required the use of some kind of shorthand notation, which would then itself need to be transcribed into standard text—this was Bartleby's job back in the office according to the quote above. This sort of two stage process—initial compact transcription followed by transcription in the full writing system—is something that continues to this day in legal proceedings and elsewhere, though more typically with stenotype keyboards rather than shorthand. The fastest operators of such machines achieve much faster transcription speeds than are possible with conventional keyboards, so such methods really are in service of speed.

Unlike watching someone write on a grain of rice, people using shorthand or stenotype machines are unlikely to draw paying spectators. Even today, however, there will be some demand for people with such skills in specialized circumstances such as legal transcriptions. Absent professional necessity, however, what might make one adopt such methods for one's own text entry needs? For many, the only text entry that they perform is for their own communication needs, so the key question would be whether their available methods are fast enough to keep up with their own language production. For example, to write this current paragraph, the bottleneck is not really the speed of typing on a QWERTY keyboard, rather the composition process itself,[3] i.e., the specific text entry mechanism that is being used is fast enough for the required work, thus providing little motivation to learn to use a stenotype keyboard.

That might not be the case if this were being typed on a mobile phone while standing on public transport of some sort, hence restricted to a single hand to hold and manipulate the device. In that case, perhaps the demands of text entry—however it is being accomplished, e.g., thumb typing or gesturing—might slow the rate down to a point where it becomes the bottleneck rather than the composition. One reaction to such a circumstance might be to defer writing the document until such time as a more suitable text entry method is available—wait until one is back at the regular keyboard to write. If that is not an option, then maybe one can live with the slowness and just deal with it; or one might learn a new method that allows for faster text entry.

While we all may reach a point where we resolve to learn a new technology to improve our text entry speed (even without a livelihood-impacting need),

---

[3] See Section 8.2 for discussion of models of the writing process, including composition.

individuals will have a particular incentive to find faster methods if they are
(1) dependent on text entry for their communication needs; and (2) have
text entry speeds that profoundly hinder their ability to communicate. These
latter conditions hold for a great many users of Augmentative and Alternative
Communication (AAC), which we touched upon in Chapter 5. When relying
on text entry for face-to-face conversation using a method that yields less
than 10 words per minute, it becomes very difficult to have the kinds of
spontaneous reciprocal conversations that many of us take for granted. So,
while many people these days are content to stick to QWERTY keyboards
as 'fast enough', there are some faced with quality of life impacts of slow
typing (AAC users or those within specialized professions) who may remain
motivated to learn faster methods, which are the topic of this chapter.

We begin examining such speed-impacting methods by looking at spontaneous and systematic methods for writing more compactly in general, as well
as methods permitting more compact gestures. Later in the chapter, we will
examine the potential of generative models in assisting with faster text entry, either by expanding/converting non-standard text into the conventional
writing system, or by enabling fast production of well-formed text.

6.2 Compact writing and/or gesturing methods

One way to speed writing is to simply write less. Provided the resulting text
encodes the same content, then why not? Well, one possible reason not to
write less is interpretability—can the message recipient read and interpret
it correctly? If the shorter encoding leaves out some important part of the
content, then communication is impacted. Still, shorter text is often perfectly
interpretable (possibly with some training), so these are important methods;
and later in the chapter we will examine methods to automatically expand
such compact writing to the full intended text.

6.2.1 Abbreviation

Consider the following request:

        Plz dnt tell me Snst Blvd's ur fave B. Wldr film

The person making this request[4] is heavily abbreviating the text relative
to standard English spelling. This aggressive abbreviation could be due to
a number of factors: to increase the speed (or reduce the effort) of typing;
to conform to specific medium restrictions on (or extra cost due to) lengths

---

[4] Probably because everyone knows that Double Indemnity is Billy Wilder's best film.

of messages (e.g., teletype or SMS messages); or even to signal an informal register in discourse.

This particular example contains some abbreviations that are conventional and some that are not, and tokens were abbreviated using different techniques. A few were not abbreviated at all—typically short words (*tell, me, film*) where any intelligible abbreviation would not yield much space savings. "Blvd" (boulevard) and "fave" are examples of conventional abbreviation— *Blvd.* is a standard form that is commonly used in addresses and maps, and *fave* is a common abbreviation of *favorite*. Both are found in dictionaries, and they would be recognizable to English readers even as stand-alone words, without any context to help understand what was intended. Other abbreviations in the example were spontaneous ad hoc abbreviations, with some letters deleted or changed to allow interpretation in context. These do not appear in dictionaries and might not be recognized correctly if presented in isolation. Sufficient context may be present within a sentence or as part of a dialogue to interpret them more-or-less easily.

Most of the spontaneous abbreviations in the example simply remove letters from the word, usually vowels. "Snst" (*Sunset*) and "Wldr" (*Wilder*) delete all and only vowels from the words, which is an extremely common approach. This is also how "dnt" (*don't*) is formed, with the additional deletion of the apostrophe indicating that the word being abbreviated is itself a contraction. Wilder's first name, Billy, is abbreviated as a prefix "B." which is generally a common method of abbreviation (e.g., the conventional abbreviation Ave. is a prefix of the intended word, *Avenue*) and is also quite common in particular as a single letter abbreviation for proper names.

The two abbreviations in the example that we have yet to address have more going on than just deletion. "Plz" (*please*) does indeed delete all of (and only) the vowels in the word, but additionally changes 's' to 'z' to indicate the voicing at the end of the word. Finally, "ur" (*your*) could be seen as deleting the first two letters of the word, but in fact this works as an abbreviation because the pronunciation of the letter 'u' is identical to *you*, thus standing in for that part of the word. Another example of this sort of pronunciation-based substitution is the use of the digit '2' in abbreviations such as '2day' for *today*.

As far as understandability of the example, the proper names (*Sunset Boulevard* and *Billy Wilder*) would be most likely to cause difficulty. "Snst" is helped by being adjacent to "Blvd", since Sunset Boulevard is both a well-known location in Los Angeles, CA and the title of an Oscar-winning film 75 years ago. "Wldr" may be understood as Wilder even for those who do not know of Billy Wilder or associate him with the film, since there are so few other tokens that would correspond to that particular abbreviation, but associating "B. Wldr" to the full name almost certainly requires context for most people. That context could come from Sunset Boulevard or from prior exchanges in a dialogue, but the message creator would need to be confident that the intent of that particular abbreviation was clear to the recipient.

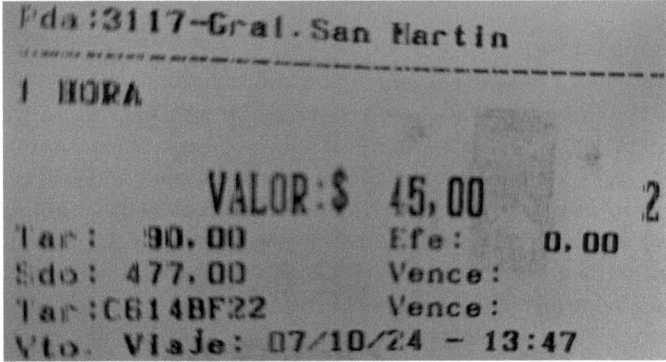

Fig. 6.1 Portion of a printed bus ticket in Montevideo, Uruguay. Source: own work.

The example above hardly exhausts the mechanisms that are used for spontaneous abbreviation. For example, acronyms (e.g., *UNICEF*) and initialisms (*FBI*) use the first letters of multiple words to form a single token, which is pronounced as a word or spelled out, respectively, when spoken. In addition to proper names, common phrases can be abbreviated in this way, such as *MYOB* or *YOLO*.[5] More than a single letter can be combined, such as *FedEx*—and like acronyms and initialisms, such methods are particularly common for names (locations, organizations, etc.), and can often become conventional.

Different languages may differ in common patterns of abbreviations used. For example, Figure 6.1 presents a portion of a printed bus ticket in Montevideo, Uruguay, which—presumably due to space constraints in the layout of the ticket—contains multiple abbreviations. Interestingly, the most common method (4 out of 7 abbreviations on the ticket) is to delete letters other than the initial letter and the final syllable, including: "**Pda**" (*parada*, meaning 'stop', as in 'bus stop'); "**Gral**" (*General*, the military rank of San Martin, the namesake of the bus stop's street); "**Sdo**" (*saldo*, 'balance'); and "**Vto**" (*vencimiento*, 'expiration'). Given word-final gender marking in Spanish, the last syllable often has particular utility in identifying the word, which explains the relatively frequent retention of this part of the word versus typical ad hoc abbreviation in English. The other three abbreviations used word prefixes, which is also common in English: "**Tar**", which is used twice for both *tarifa* ('rate') and *tarjeta* ('card ID'); and "**Efe**" for *efectivo* ('cash'). The use of the same abbreviation for two different words emphasizes the utility of context for understanding the intended word, since the values associated with these fields only make sense under particular interpretations.

Different scripts also impact the kinds of abbreviatory mechanisms that are used for spontaneous abbreviations. Gorman & Roark (2024) presents a survey of abbreviation methods used in 50+ languages, covering a diversity

---

[5] *MYOB* stands for "Mind your own business"; and *YOLO* means "You only live once."

of language families and scripts. Languages using the Perso-Arabic script, the writing systems of which already typically omit vowels, are shown to make use of relatively few methods of abbreviation. Nor do languages using Brahmic scripts, largely because methods for indicating deletion of vowels in these scripts often require extra markings, i.e., they are not shortenings at all. Interestingly, many abbreviations of names in India using first character abbreviations are based on Latin script first characters, such as the political party "BJP", written in Hindi as बीजेपी, which sounds out the letters of the Latin script initialism (bee-jay-pee). Other alphabets, such as Cyrillic, have many of the same methods for abbreviation as the Latin script, and there are some attested methods in languages surveyed in Gorman & Roark (2024) that are not typically found in English.

While some abbreviations become conventionalized and are even included in dictionaries, many of these methods are used spontaneously by people to avoid writing out the full words. Sometimes these are intended to be shared with others directly, and can be intended (as mentioned above) to signal informality in discourse (e.g., on social media).[6] Sometimes, however, it is simply used as an expedient, and individuals would prefer to have the message encoded with correct orthography. In those cases, there are systems for expanding spontaneously abbreviated text, which we will discuss later in the chapter.

In addition to spontaneous abbreviation that we have discussed in this section, there are writing methods based on well-defined systems that are intended to achieve the same ends, but without the potential for misunderstanding that exists for ad hoc abbreviation. Such systems are known as shorthand.

6.2.2 Shorthand systems

Shorthand systems for a language are designed to allow for ease and speed of transcription rather than ease and speed of reading, and are most often intended for quick, temporary notes—something of a memory aid—rather than as a final transcription. The common process (as hinted in the quote from Melville's story at the beginning of the chapter) is to transcribe in shorthand, then copy into standard orthography afterwards. Such systems have been defined since antiquity (e.g., Tironian shorthand, used to transcribe the speeches and dictation of Cicero), and many competing systems were popularized beginning in the 17th century in Europe and, later, in North America.

---

[6] Non-standard spellings that are not abbreviations can also serve the purpose of signaling informality or some other extra-linguistic information, such as "Coooooooooool".

The Alphabet of Gregg Shorthand

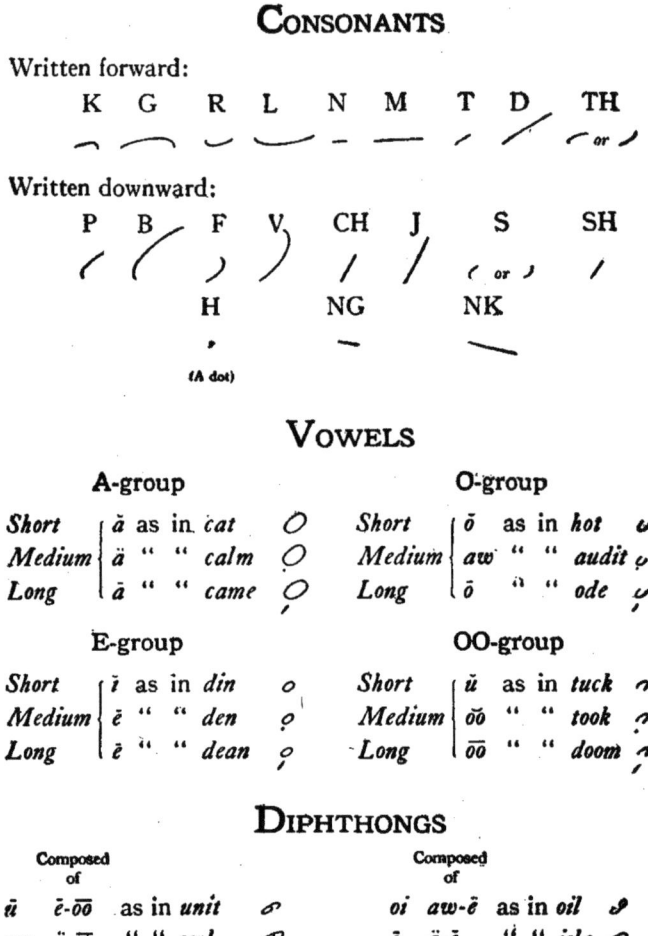

CONSONANTS

Written forward:

K G R L N M T D TH

Written downward:

P B F V CH J S SH

H NG NK

(A dot)

VOWELS

A-group

Short	*ă* as in *cat*	𝒪
Medium	*ä* " " *calm*	𝒪
Long	*ā* " " *came*	𝒪

O-group

Short	*ŏ* as in *hot*	⌣
Medium	*aw* " " *audit*	⌣
Long	*ō* " " *ode*	⌣

E-group

Short	*ĭ* as in *din*	o
Medium	*ĕ* " " *den*	ọ
Long	*ē* " " *dean*	ọ

OO-group

Short	*ŭ* as in *tuck*	∩
Medium	*ŏŏ* " " *took*	∩
Long	*ōō* " " *doom*	∩

DIPHTHONGS

Composed of			
ū	*ē-ōō*	as in *unit*	ℐ
ow	*ä-ōō*	" " *owl*	ℐ

Composed of			
oi	*aw-ē*	as in *oil*	ℐ
ī	*ä-ē*	" " *isle*	ℴ

Fig. 6.2 Basic alphabet of Gregg shorthand, as presented in Gregg (1916). Work is in the public domain in its country of origin and other countries and areas where the copyright term is the author's life plus 70 years or fewer.

As an example,[7] we will briefly present Gregg shorthand, originally designed for English in the late 19th century and subsequently adapted to many other languages. Gregg shorthand was developed as an alternative to the already popular Pitman shorthand, which was also a phonemic system, i.e., a method for transcribing the phonemes being pronounced rather than

---

[7] See Daniels & Bright (1996) for further history of shorthand systems.

GENERAL EXERCISE

knee	n ē		**tact**	t ă k t	
keen	k ē n		**tray**	t r ā	
kick	k ĭ k		**train**	t r ā n	

Fig. 6.3 Some Gregg shorthand word transcription exercises from Gregg (1916). Work is in the public domain in its country of origin and other countries and areas where the copyright term is the author's life plus 70 years or fewer.

approximating the writing system. Pitman made certain distinctions, e.g., voicing of consonants, by the thickness of the lines,[8] whereas Gregg made such distinctions by the length of the stroke. Figure 6.2 presents a table of the basic alphabet used in the shorthand, as presented in Gregg (1916). The difference between a 't' and a 'd' is that the latter is voiced and the former is unvoiced, hence the difference in the strokes between these two conso-nants is simply length. The same distinction can be found between other voiced/unvoiced pairs such as k/g, p/b and f/v, as well as between other similarly pronounced consonants as r/l and m/n. Differences in vowel length are marked with dots or lines underneath the vowel strokes.

Unlike Pitman shorthand—which, as with ad hoc abbreviations discussed earlier, permitted omission of vowels if they were unneeded to distinguish the word—Gregg shorthand required inclusion of the vowels, but they were designed to be relatively easy to incorporate into a stroke. Figure 6.3 shows several example word transcriptions using the system. The silent 'k' in the word *knee* is not included in the Gregg transcription, just the 'n' consonant and the long vowel from the E-group, which is indicated with a small loop and a line underneath. The short vowel from the E-group, as in the word *kick*, just requires a small loop without a line underneath. In that example, the 'k' is not silent, hence must be transcribed.

Beyond the compact strokes, Gregg shorthand also included signs for com-mon full words and phrases, as well as for common suffixes such as *-tion*. Many other time/space saving methods are included in the system, such as conjoin-ing phrases without spaces between the words, and omission of some words from common constructions, such as the preposition in common phrases such as 'day by day' or 'day after day'. Additionally, word prefixes can be used as abbreviations, often dictated by whether the context is sufficient to disam-biguate (Gregg, 1916, , p.62):

It is important to bear in mind that all the words so abbreviated will usually occur in sentences. For instance in the sentence "He was received with great enthusiasm,"

---

[8] Recall that Pitman shorthand was influential in the creation of the Swampy Cree script, which used similar distinctions, as discussed in Section 3.4.

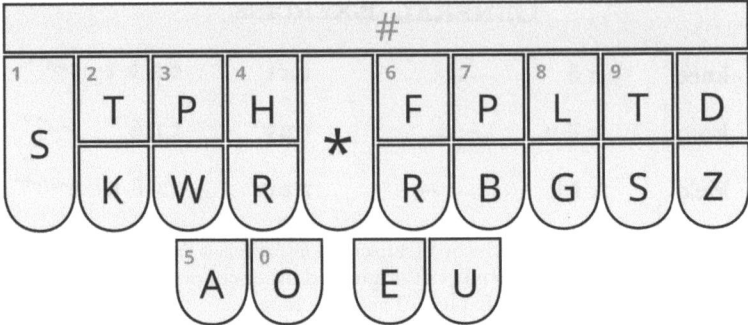

Fig. 6.4 Common layout of an English stenotype keyboard. Image licence CC BY-SA 4.0, `https://en.wikipedia.org/wiki/File:Stenotype_en_layout.svg`

> it would be sufficient to write *enthus* for *enthusiasm*; and the same form might be used for *enthusiastic* in "He met with a most enthusiastic reception."

The last update[9] to the Gregg shorthand system was in 1988, and the system continues to have some users, particularly those who require fast note-taking in the field, without access to higher technology methods. In more controlled settings, such as courtrooms, use of stenotype keyboards provides a speedy alternative to such handwritten methods.

6.2.3 Stenotype keyboards

Stenotype keyboards have their origins in mechanical devices that permitted a kind of typewritten shorthand, which was then converted into the standard writing system in a second pass, much as the kinds of handwritten shorthand systems described above are. Unlike handwritten shorthand, however, which requires specialized optical character recognition systems to convert to machine-readable text, stenotype keyboards can straightforwardly be adapted for use as computer keyboards, thus permitting application of automatic conversion software to yield full text without requiring manual transcription. After extensive training, such systems yield typing speeds far faster than standard QWERTY-keyboard touch typing, sufficient to transcribe human speech in real time. Figure 6.4 presents a diagram of a common layout of a stenotype keyboard for English.

As alluded to in Chapter 5, stenotype keyboards change some important assumptions built into conventional typing, primarily via multiple keys being pressed simultaneously at each step. For this reason, such keyboards are often referred to as chorded keyboards, since the multiple key gestures are similar

---

[9] `https://greggshorthand.github.io/abcenten.html`

to what are used on a piano to produce a chord. Much like the Decapoint (Braille, 1839) system that was implemented in the Raphigraphe machine in 1843 (see Figure 5.7 in Chapter 5), mechanical stenotype machines associated each key on the keyboard with a particular printing location. The Raphigraphe produced columns of dots/pixels to allow a visual representation of a letter to emerge over multiple columns. Stenotype machines instead produced a row of letters, with the same number of columns as the number of keys on the keyboard. In modern times, for English, the conventional keyboard and column layout included the following 22 letters in the following positions:

STKPWHRAO*EUFRPBLGTSDZ

From this we can note that some letters (like S) are present multiple times, and some letters in the English alphabet (like C, B, M and N) are not present at all. When keys are pressed in the keyboard, a row showing all of the pressed letters is produced on the paper, and the paper is advanced to the next row. For example, to type the word *held*, the keys for H, E, L and D would be depressed simultaneously on the keyboard, and the resulting row would show:

H   E   L   D

Obviously, standard orthography is not presented in this way, so the printed result would then be transcribed—something that is now typically done automatically.

The chording of the word is responsible for some of the speedup that can be obtained with such keyboards—allowing for a single multi-key action rather than four separate, sequential single key actions (plus space). Much of the speedup, however, is due to the use of the specific transcription standard, known as a 'theory', that is followed, as well as the establishment of abbreviations, known as 'briefs' in the stenotype community. Unlike the above example of *held*, most words would be transcribed distinctly from the standard orthography. As with shorthand systems discussed earlier, most theories use phonological methods to transcribe the words, and missing letters are transcribed by convention using other letter combinations. For example, 'B' may be transcribed as 'PW' in the particular theory, though theories can (and do) differ in their conventions. Understanding which conventions are used is key to being able to transcribe them afterwards. Further, the set of briefs (abbreviations) can be particularly idiosyncratic, something specifically designed by the user of the system, perhaps with a particular common topic in mind. For example, if the stenotype is being used for legal transcription, then many common legal words and phrases could be included in the system as briefs for speed.

Given the impressive speedups that are achieved by professionals using such systems, one wonders why one does not see more widespread adoption of the approach for general text entry. Learning curve is certainly one reason—mastering a theory and establishing the set of abbreviations is neither fast nor easy. As alluded to at the beginning of the chapter, another reason for many people is likely that current text entry methods are fast enough, since the bottleneck can often occur in composition rather than typing.

Additionally, for those for whom typing is very slow, such as those using AAC technology, who might otherwise be willing to put in the time to learn a much faster method of text entry, chording as a gesture may not be particularly accessible. However, we do see the establishment of personalized pre-stored word, phrase and even full sentence lexicons for individuals in many AAC speech generating devices (see Section 5.5.1). These items can be selected within the interface, using the input gestures configured for the individual, in ways that allow for easier access than fully typing them out, much as with 'briefs' in a stenotype keyboard. In such a way, some of the speedups attributed to stenotype keyboards are achieved in these sorts of specialized systems.

6.2.4 Romanization

For some languages using non-alphabetic writing systems, direct keyboard-driven text entry is difficult, as described in Chapter 3. For that reason, romanization—writing in the Latin script—is often employed for its relative ease of text entry. The romanization can be part of a formally defined system, as in pinyin input for Chinese, or spontaneous ad hoc romanization, which is common in India and other countries of South and Southeast Asia. Here we will describe a couple of text input methods based on romanization, for which a conventional QWERTY keyboard (physical or virtual) is appropriate.

The Hanyu Pinyin romanization system has been the standard of the Chinese government (among others) for Mandarin Chinese since the 1950s, and it subsequently became the basis for the most common text input method for Chinese as well. Text input using pinyin makes use of an input method editor (IME), that acts as an intermediary between the keyboard and the ultimate destination of the text, such as a document. Latin script text is input, and the IME presents target output text options that are potentially intended given the input. For example, if one inputs "ji", then the IME may present the following five options: 1. 及 2. 级 3. 几 4. 即   5. 机. To select the desired output, the number of the selection is typed (or space bar if the top ranked one is correct). Each of these five tokens are romanized as "ji",[10] which is based on their pronunciation. Note, however, that the given romanization ignores the tone, which differs for several of the options. The Hanyu Pinyin romanization system upon which the IME is based does include diacritics for tone, but diacritics are typically omitted to speed up input.

---

[10] As stand-alone words, these translate to "and" (及), "class" (级), "several" (几), "immediately" (即) and "machine" (机).

Ranking of the possible outputs is based on the probability of each output string given what has already been typed. This will overall make the system much more accurate in its rankings—i.e., the top ranked one is more likely to be the desired output—though it means that, for the same romanized input, what is top ranked will vary, thus generally requiring the user to attend to the options as opposed to having memorized common rankings and selecting without looking. If the desired output is not available, an IME will also provide an option to scroll through more possible options.

While the above description illustrates the basic functionality of the input method, modern IME systems include many optimizations to speed up entry. First, one does not need to make a selection on a character-by-character basis; rather, one can enter multiple characters before selection, thus providing even more context to the predictions being made by the system, and having to attend to the rankings less frequently. Second, as with shorthand and stenotype systems, abbreviations are typically supported, both user provided and conventional. Finally, as with many text-entry systems, personalization can influence what the predictive model thinks is likely, and next-word prediction can go beyond what has already been typed to suggest what might be coming next. Similar romanized input methods exist for Japanese, based on different long-standing romanization standards—though as noted in previous chapters, many Japanese speakers type using Hiragana, where the problem of converting to Kanji is still the same.

Romanized text entry in South Asia—India and surrounding countries—is also common, though without the benefit of widely-accepted (or widely-used) romanization systems. While such romanization systems do exist for languages written in Brahmic scripts, such as the ISO 15919 standard,[11] they are not commonly used by native speakers when spontaneously writing their languages in the Latin script—something that is, for historical reasons, extremely common in the region. Many individuals simply choose to use the Latin script to write their native languages in many scenarios, despite the fact that there is no orthography in these languages in the Latin script, leading to high levels of spelling variability (Brandt, 2020). Frequent code-switching with English and other regional languages, along with many cross-lingual loan words, complicate the picture.

Romanized Hindi, for example, can be thought of as a rough phonetic transcription using the Latin script. Wolf-Sonkin et al. (2019) provide an example taken from the comments of a blog entry that begins: "Bhrashtachar aam aadmi se chalu hota hai..." which would be written in the Devanagari script as भ्रष्टाचार आम आदमी से चालू होता है... and which they roughly translate to "Corruption starts from the common man..." They note that "bhrashtachar" in particular is a word that shows up many times in the comments of that particular article, with many different spellings. The intent of the Wolf-Sonkin et al. (2019) paper was to advocate for the use of transliteration within the

---

[11] https://www.iso.org/standard/28333.html

predictive model used for text entry, while leaving the output text in the Latin script, as the means of (system internally) normalizing many possible spellings of the same word in the Latin script to a single spelling in the Brahmic script.

Beyond just leaving the text in the Latin script, there are keyboards that will transliterate text input in the Latin script into the native script of the language (Hellsten et al., 2017). This is accomplished by integrating methods to convert input from a QWERTY style Latin script keyboard to likely transliterations within existing mobile keyboard systems. In contrast to the pinyin input methods discussed earlier in this section, the handling of the high levels of spelling variability in this scenario adds to the overall uncertainty inherent in the input method, which can have an impact on the speed of text entry. It does, however, provide a useful alternative to text entry either purely in the native script of the language (typically difficult to represent for Brahmic scripts on small mobile keyboards—see Chapter 3 for discussion of similar issues even for larger typewriters) or purely in the Latin script (not the native writing system of the language).

This latter example is a case of converting the input text as it is being typed, which is very common as it provides immediate visual feedback to the individual who is typing, for monitoring and possible correction. Another alternative is for the user to produce some text, then submit that text as a whole for conversion. This is often how document translation occurs, rather than word-by-word while the original is being created. It is also the approach in the original paradigm of shorthand and stenotype transcription. We will next briefly explore the scenario of these sorts of off-line processes.

6.3 Automatic conversion of compact text

For this section, we will focus on automatic conversion of text after composition, at the document, paragraph or sentence level. This kind of off-line post-hoc processing stands in contrast to on-line word-by-word (or even letter-by-letter) processing during composition, a distinction we touched on in Section 4.3.1. The on-line use of predictive models during composition is common during text entry on mobile devices, a scenario where errors are typically more frequent and auto-correction is often required, something return to later in the chapter. Off-line processing of whole sentences or documents has a long history in machine-readable document preparation, such as when documents are "passed through a spell-checker". Several methods of compact writing and/or gesturing that we have covered in this chapter are well-suited to this paradigm of post-processing, including conversion of ad hoc abbreviations or stenotype keyboard output to fully expanded orthography, as well as transliteration of spontaneously romanized text into other writing systems.

The use of spell-checking software has been commonplace for decades, as discussed in Chapter 4, with a common approach being identification (through highlighting or underlining) of potentially misspelled words in a document, and suggestions of alternatives for these words, which can be selected or ignored. Terms can even be added to a personalized dictionary to avoid subsequent flagging. Depending on the certainty of the system, a spelling error could be auto-corrected without requiring confirmation, but this is a system design issue and may depend on the specific interface being used—again, common on mobile devices, less so for other document processing scenarios. For example, creation of a longer document on a laptop computer may involve less attention to the correctness of the produced text immediately during composition and more post-hoc checking of the entire document.

Consider the following excerpt from the 1994 science fiction novel *Feersum Endjinn* by Iain M. Banks, from which a sentence was included in the analysis in Chapter 4. In the novel, one of the characters, Bascule, keeps a journal that includes the following observation:

> ...I cant spel rite, juss 1/2 2 do evrythin foneticly. Iss not a problim cos u can put eny old rubish thru practikly anyfin evin a chile's toy computir & get it 2 cum out speld perfictly & gramatisized 2 & evin improvd 2 thi poynt whare yood fink u waz Bill bleedin Shaikspir by thi langwidje.[12]

This is a relatively recent example of eye dialect, the literary use of non-standard spelling to (typically) indicate a non-standard accent and/or lack of literacy in the individual, a device notably used by, among others, Mark Twain and Charles Dickens in the 19th century. Based on the above sample, we can glean something about Bascule's intended accent (e.g., from spelling the word 'think' as *fink*). But the content of the observation—that computer systems could clean up and improve badly spelled and/or poorly written input text—is no longer science fiction.[13]

To casually reinforce the more thorough results presented in Chapter 4, we passed the above full paragraph through two commercially available web

---

[12] From part 4 of chapter 3 of *Feersum Endjinn*. Transcription into standard English orthography: "...I can't spell right, just have to do everything phonetically. It's not a problem because you can put any old rubbish through practically anything even a child's toy computer and get it to come out spelled perfectly and grammatisized too and even improved to the point where you'd think you was Bill bleeding Shakespeare by the language."

[13] Also, if it was true that textual clean up was readily available in Bascule's time, why had the journal not been so cleaned prior to presentation to the reader? It seems that Banks had more stamina than, say, George Bernard Shaw, who, in Pygmalion, gave up a similar effort at the end of Eliza Doolittle's third line, as follows: "Ow, eez ye-ooa san, is e? Wal, fewd dan y' de-ooty bawmz a mather should, eed now bettern to spawl a pore gel's flahrzn than ran awy atbaht pyin. Will ye-oo py me f'them? [Here, with apologies, this desperate attempt to represent her dialect without a phonetic alphabet must be abandoned as unintelligible outside London.]" Granted, Bascule's text is quite a bit easier to understand than Shaw's, outside of London at least.

interfaces to large language models (LLMs) with the following instructions: "The given paragraph is English with irregular spelling. Can you rewrite it with standard English spelling? Here is the paragraph:" followed by the above quote from Feersum Endjinn. Both produced correct English spelling for the entire passage, one with a word-by-word correspondence and one with a slight alteration to one part of the paragraph. For the portion of the passage "...gramatisized 2 & evin improvd..." one LLM produced "grammatized too, and even improved", while the other produced "grammatically correct". This was the only difference between the outputs of the two LLM chat interfaces for this prompt. The latter system accurately paraphrased "speld perfictly & gramatisized 2" as "spelled perfectly and grammatically correct" but left out the "evin improvd" part.

Within a paradigm where one, like Bascule, simply composes a sentence, paragraph or full document without worrying about orthographic correctness, there are now many powerful tools available to assist with subsequent transcription and/or editing stages. Consider the use of abbreviation discussed earlier in this chapter, which can provide important text entry speedups for those for whom selecting each letter in a word is slow or effortful, such as those using AAC systems. The idea of supporting abbreviation in such systems has a long history. Early efforts to provide speedups for individuals using speech generating devices via abbreviation were focused on methods for including abbreviations in an individual's system vocabulary (e.g., Kelso & Vanderheiden, 1982; Vanderheiden & Kelso, 1987), much like 'briefs' are used with a stenotype keyboard. This requires that the individual memorize each specifically included abbreviation, which limits the number that can be included in the system. Later work (e.g., Demasco et al., 1989; Stum & Demasco, 1992) looked to allow for spontaneous abbreviations, with possible expansions identified and disambiguated using predictive models. Shieber & Nelken (2007) also allowed spontaneous abbreviations, but additionally made use of a stipulated compression method, which in their case (for English) involved removing non-initial vowels and doubled consonants, so that, e.g., the word "aggressive" would be abbreviated as "agrsv".

Cai et al. (2024) provide a recent example of this long-standing idea, specifically focused on individuals with ALS using eyegaze direct selection methods for typing, but with a relatively extreme stipulated compression method: simply type the first letter of each word in the intended sentence. This is not a word-by-word strategy, i.e., one does not type one letter and then look for the possible expansions of that letter. Rather one enters the string of initial letters, then the system provides possible full sentences that match the input. The system described in their paper makes use of dialogue context (e.g., previous utterances in the conversation) and the already entered initial letters to elicit possible full sentences from a large language model (LLM) for the user to select. To make the system fully functional, it additionally provides LLM-driven utilities for eliciting alternatives to specific words in the suggested sentences and/or for generating more alternatives. In a study that included individuals with ALS using the system both in the lab and in the field, they found significant reductions in keystrokes required to perform text entry using this system over the baseline.

We can briefly illustrate how the system works, as well as potential challenges, based on a manufactured example. Suppose someone asks "How are you feeling today?" and in response the individual intends to reply "I don't feel very well today" and types "idfvwt". The system makes use of the original question—perhaps via automatic speech recognition—and the six letters as context, and returns a set of candidate sentences for the individual to select. We simulated this with a couple of commercial LLMs with the following prompt: "suppose you are generating a response to the following question: 'How are you feeling today?' The response should have 6 words and the first letters of the 6 words should be i, d, f, v, w, t. Can you give five alternative possible responses?" This is how a system such as the one described in Cai et al. (2024) could be implemented: when the system user submits their abbreviated sentence (in this case "idfvwt"), the system could submit a query to one or more LLMs that includes explicit instructions along the lines of what was provided above. In this case, our responses included:

> I don't feel very worried today.
> I'm doing fine, very well today.
> I've decided fully, very well today.
> I do feel very warm today.
> I'm definitely feeling very well today.
> I'm delighted, feeling vibrant, wonderfully tranquil.

Unfortunately, the intended utterance was not included, although the first suggestion is just one word off. In the system described in Cai et al. (2024), the user could select the closest example, then select the specific word to edit (in this case "worried") at which point alternatives beginning with 'w' would be provided, hopefully including the intended one.

Examining the outputs of the LLMs above, it appears that there is a strong bias towards positivity, which perhaps is why the intended response was missing. Cai et al. (2024) make use of fine-tuned LLMs, meaning that the pretrained models were further trained to specifically perform this abbreviation expansion task, in contrast to just being given a prompt that tells the model what to do. In addition, for the experiments reported in their paper, they turned off personalization of the text entry systems, so that the system did not have access to any personalized vocabulary or prior usage patterns— just the conversation context. The integration of such general purpose models with personalized systems is a challenge we will raise later in the chapter.

Conversion of sentences or full documents from the one script to another script would also fit this paradigm, thus addressing the romanization scenarios presented earlier in the chapter. As with the above abbreviation example, a system designed to perform this task would be expected to provide conversions that directly correspond to the input. For abbreviation expansion, if there are 6 word initial letters, then the suggested expansions should have exactly six words, all starting with the appropriate letters. For transliteration, the resulting text should correspond to the same linguistic content. In other words, if an individual can read both scripts, when they read the text aloud it should be the same, regardless of which version they read. Ensuring that the system follows such hard constraints on the task can also be a challenge when making use of LLMs.

Now that we have discussed off-line conversion of already composed text, we can turn to some scenarios that involve conversion during composition, which invariably involve the use of predictive models.

6.4 Use of predictive models during composition

We have chosen to distinguish between conversion to standard orthography of completed compositions — sentences or even documents—and on-line modeling used during composition, either word-by-word or even letter-by-letter, which we cover in this section. Instead of building systems to convert from one representation to another, such on-line systems instead rely on language models to assign probabilities, given what has already been entered, to distinct potentially intended letters, words or phrases, which are presented for selection (or sometimes automatically selected) within the text entry interface. We have already seen some examples of the use of predictive models during text composition, as part of romanized input systems such as pinyin. Here we will go a bit deeper into what kinds of models are used for different kinds of input functionality, beginning with some of the most ubiquitous in contemporary mobile keyboards.

6.4.1 Next-word prediction, word completion and auto-correction

Predicting a missing word in a sentence is a very common objective when training even the largest language models, an approach which has been responsible for impressive advances in AI in recent years. Generative models make such predictions based only on the words that preceded the word to be predicted, which is exactly the next-word prediction functionality used to provide suggestions to users within a dynamic region of a virtual keyboard. For example, if an individual types "What is your ", then likely next words may include "name", "problem", "birthday", "biggest" and so on. A model is trained to assign probabilities to such continuations, and the most likely continuation can be included as a suggestion.

Continuing with the above example, if, after typing "What is your ", the individual then types the letter 'b', then some of the likely next words become much less likely,[14] while words such as "birthday" and "biggest" presumably become more likely. The same model that was used for the earlier next-word candidates can provide updated probabilities given this new information. However, for a number of reasons, once typing of the current word begins, this is generally considered a different task from next-word prediction, known as word completion. The most critical distinction between next-word prediction and word completion is that, for the latter, the system must account for the lack of certainty of what letter was actually intended.

When using a physical QWERTY keyboard, such as that attached to a desktop or included in a laptop computer, pressing the key associated with a letter types that letter. This happens via a switch attached to each key that is either activated (switched 'on' when being pressed) or not ('off' when not pressed). Each key is either on or off.[15] On a virtual keyboard, such as those commonly used on mobile devices, there is no physical key; rather, there is a region on the screen associated with each letter. The line surrounding each letter's region is its bounding box, and one can achieve something similar to a mechanical keyboard's binary on and off functionality by counting a touch inside a bounding box as 'on', otherwise off. For convenience, we can refer to the string of letters that would be typed under such an approach as the 'literal'. For example, if one intends to type the word 'slowly', one might produce the literal string 'slowky' if the actual point touched by the finger—the *touchpoint*—for the second letter 'l' strays into the bounding box of 'k', as in the schematic in Figure 6.5, which shows each of 6 hypothetical touchpoints. Such divergences between the intended and literal strings is not uncommon; given the relatively small sizes of each letter's region on these virtual keyboards, the touchpoints when intending the letter will vary. Sometimes a touchpoint may be right at the center of the letter's allocated region, as with touchpoint 2 in Figure 6.5; other times it may fall farther away from the center. The question is how a text entry system can infer the intended string when such minor touchpoint divergences are common.

To explain how this may work, we will use some math, for which we will set up some relatively simple notation. For ease of exposition, suppose we are speaking of single intended words consisting of letters that come from a fixed set of letters Σ, such as the word 'slowly' consisting of letters that come from $\Sigma = \{a - z\}$, the set of Latin script letters from 'a' to 'z'. Let $\mathcal{S} = s_1 \ldots s_m$ be the intended typed string of m letters, where all s_i are drawn from the set Σ, and $\mathcal{T} = t_1 \ldots t_n$ the input string of n touchpoints.[16] Thus, for the above slowly/slowky example, $s_5 = 1$ and t_5 is a touchpoint somewhere within

---

[14] While words that do not start with 'b' will have lower probability after 'b' is typed, their probability will not really drop to zero, since it is possible that the letter was typed in error.

[15] Readers may have had the experience of a key being sticky and getting stuck in the on position, which can lead to lettttttter repetttttittttions.

[16] The number of touchpoints may differ from the number of letters in the intended word, if too many or too few points are touched by the user.

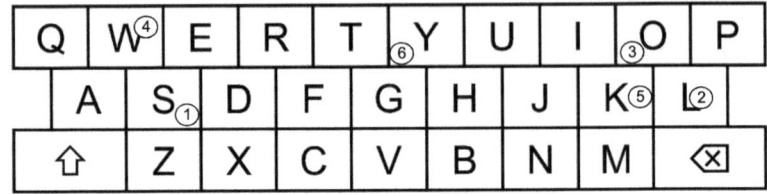

Fig. 6.5 Numbered touchpoints on a virtual QWERTY keyboard for the target word
`slowly`. Source: own work.

the bounding box of the key associated with 'k'. Given these touchpoints $\mathcal{T}$,
a text entry system will return the most likely intended string $\widehat{S}$:

$$\widehat{S} = \underset{S \in \Sigma^+}{\operatorname{argmax}} \, \mathrm{P}(S \mid \mathcal{T})$$

i.e., the word S consisting of one or more letters (Σ^+) that has the highest
probability conditioned on the input touchpoints $\mathrm{P}(S \mid \mathcal{T})$. There are (at
least) two key challenges with this formulation of the problem: first, how
to estimate the conditional probability P of each possible word S given the
touchpoints $\mathcal{T}$; and second, how to efficiently explore the space of possible
words to find the one with the highest probability.

As far as estimating the probability, this can be accomplished in a variety
of ways. Mapping from a sequence of touchpoints to a sequence of letters
(input touchpoints, output text) is broadly speaking a sequence-to-sequence
task similar to machine translation (input text in one language, output text
in another) or speech recognition (input speech, output text), hence methods
can be pursued that are similar to methods used for those tasks, such as
neural networks. Both machine translation and speech recognition, however,
have a key advantage over text entry when training models: data is relatively
easy to obtain to train such models. Again, broadly speaking, end-to-end
machine translation models are trained largely by being exposed to many
valid translations pairing input texts with their translations; and end-to-
end speech recognition models are trained with examples of spoken language
paired with their transcriptions. The systems learn to directly take text or
speech and assign probabilities to output text conditioned on the given inputs,
i.e., they directly estimate the conditional probability shown above. To train
the same sort of end-to-end model for text entry would require collecting
many examples of input touchpoints and output intended text.

However, touchpoint/text paired data is quite sparse for a few reasons:
touchpoints are ephemeral; they are tricky to effectively (naturally) elicit;
and they are often reactions to context-specific text entry system stimuli
and/or feedback. Let us address each of these issues in turn.

Except under extraordinary circumstances—such as a controlled study—
the touchpoints are ephemeral intermediates of the text entry process that

are not retained. This is similar to certain kinds of physical keyboard gestures that are part of entering the text, but not directly reflected in the resulting text itself. For example, when typing the token `CONMEBOL` several things may or may not have happened: the caps lock key may have been used; one or more letters may have been typed in error and deleted; or the word as a whole may have been cut and paste from another document. The true sequence of input and editing gestures that were used to produce the word have been lost to history. Similarly, touchpoints are important to the process of entering the text, but once the text has been entered that information is discarded. Unlike translation and spoken language, publicly available repositories of inputs and their associated outputs do not, as far as we know, exist in any substantial amount for touchpoints and resulting text.

People may be hired to produce natural text using a device where these ephemeral touchpoints are retained, however this also has some challenges. How does one ensure naturalness and still have certainty about the intended text? Having the individual copy a given text is not a natural manner of input, even though it allows for certainty about the intended text. Producing spontaneous text can be natural—provided it is in a context and/or setting similar to typical texting behavior on a virtual keyboard; however, how does one get certainty about the intended text? One can take the resulting text as equivalent to the intended text, but mistakes may have gone uncorrected. One can ask the individual to review the resulting text on the spot, but this may interfere with the naturalness of the text entry session. After the fact, somebody (including the individual) may try to transcribe the intended text given the sequence of touchpoints. This is what is done with speech—native speakers listen to the audio and transcribe the intended words. It is also done with translation—translators read the document to be translated and translate it. Transcribing the intended text from a sequence of touchpoints, however, is a very different kind of task—possible, but difficult and unusual relative to speech transcription and translation.

A further complication for producing paired touchpoints and text is that text entry generally occurs on a mobile keyboard system that provides many dynamic utilities for the user, including next word prediction and completion, and system behavior can be informed by personalization (discussed later in the chapter). The touches are fundamentally reactions to what is presented to the user in the virtual keyboard, and hence this information would become part of what needs to be retained to fully understand the interaction, either for training a model or for transcribing the intended text. Such utilities could be turned off for data collection, but then the interaction will not be the same as with a typical mobile keyboard, hence a loss of naturalness. On balance, this is a very tricky sort of data collection to attempt, making end-to-end training from human labeled data much more difficult to achieve than for speech recognition or machine translation.

An alternative approach to estimating the conditional probability directly is to decompose it into parts that are easier to estimate independently—the so-called "noisy-channel" model, which we introduced in Chapter 4 in the context of spelling correction. Such an approach was used for speech recognition and machine translation going back as far as the 1980s—see Jurafsky & Martin (2008) or Jelinek (1998) for details—and for the task of spell-checking (closely related to inferring the intended word from the literal) in Kernighan et al. (1990). The noisy-channel model uses Bayes' rule to decompose the conditional probability:

$$P(S \mid \mathcal{T}) = \frac{P(S, \mathcal{T})}{P(\mathcal{T})} = \frac{P(\mathcal{T} \mid S)\, P(S)}{P(\mathcal{T})}$$

Further, since the touchpoints are given (i.e., the same for all competing possibly intended words), and the objective is to find the word with the maximum probability, this allows a further simplification by ignoring the (constant) denominator:

$$\hat{S} = \operatorname*{argmax}_{S \in \Sigma^+} P(S \mid \mathcal{T}) = \operatorname*{argmax}_{S \in \Sigma^+} P(\mathcal{T} \mid S)\, P(S)$$

In such a way, the original conditional probability has been decomposed into two component parts. The first $P(\mathcal{T} \mid S)$ estimates the conditional probability of the touchpoints given the word S, commonly called (for text entry) the spatial model; and the second $P(S)$ estimates the prior probability of the word S, which is the language model.

To make this concrete, consider our earlier example where the intended word is 'slowly' and the literal string is 'slowky'. An important question is whether the probability of 'slowly' is higher than the probability of 'slowky' given the touchpoints. This will depend on how the models are estimated. Ouyang et al. (2017), for example, suggest that the spatial model could be based on a neural net or simple Gaussian distribution centered at the key center, and opt for the latter in their approach. To estimate the Gaussians, they create a bi-key dependency, similar to the kind of context dependent phone distinctions made in speech recognition to account for co-articulation effects (see, e.g., Mohri et al., 2002). Much as the pronunciation of a vowel can depend on the consonants that precede or follow it, the touchpoint for a letter can depend on the letters that precede or follow it. Thus, effectively, the probability of each touchpoint in the spatial model is conditioned on the letter and preceding letter:[17]

$$P(\mathcal{T} \mid S) = \prod_{i=1}^{n} P(t_i \mid s_i, s_{i-1})$$

---

[17] While the number of touchpoints may differ from the number of letters, for the ease of exposition in the current example, we will treat them as equivalent and aligned, i.e., touchpoint t_i goes with letter s_i.

where $\prod$ means the product of each of the probabilities, and s_0 is some special beginning-of-word symbol. Note that, for the two competing words, they only differ at the fifth letter, so the difference between them in spatial probability is the difference between $P(t_5 \mid \mathtt{k}, \mathtt{w})\, P(t_6 \mid \mathtt{y}, \mathtt{k})$ and $P(t_5 \mid \mathtt{l}, \mathtt{w})\, P(t_6 \mid \mathtt{y}, \mathtt{l})$.

At its simplest, the language model $P(S)$ will simply assign high probability to common words in the language being typed (in this case English), and low probability to unlikely words. The word $\mathtt{slowly}$ is a valid and not particularly rare word in English; the word $\mathtt{slowky}$ is not a common word in English but could be a proper name, so its probability is non-zero, but clearly should be lower than $\mathtt{slowly}$. More complex modeling would look at the probability of the word in the context of its use (e.g., words previously entered), which can make a big difference.

In such an approach, the difference between the probability of $\mathtt{slowly}$ and the probability of $\mathtt{slowky}$ is:

$$P(t_5 \mid \mathtt{l}, \mathtt{w})\, P(t_6 \mid \mathtt{y}, \mathtt{l})\, P(\mathtt{slowly}) \;-\; P(t_5 \mid \mathtt{k}, \mathtt{w})\, P(t_6 \mid \mathtt{y}, \mathtt{k})\, P(\mathtt{slowky})$$

Ouyang et al. (2017) state that, if the probability of the candidate word is higher than the probability of the literal by some margin, i.e., if the above difference is large enough, then auto-correction occurs. Their approach also includes methods for dealing with uncertainty when making word completion suggestions, and a plethora of methods for decoding, i.e., exploring the space of alternative intended words given the touchpoints. This is hardly the final word, though—since larger and larger language models are being created and used. System latency is a particular issue with these keyboards—when one types a letter, one typically wants to see it show up immediately in the text buffer—so methods for making larger and more accurate language models fast enough on-device will continue to be a challenge for such systems. We return to this issue in the final section of the chapter.

6.4.2 Gesture-based input

As briefly presented in Chapter 5, gesture-based input on mobile devices involves specification of a word by tracing a path on the screen of the mobile keyboard from the first letter in the word, through every letter in the word, terminating at the last letter of the word, at which point the finger can be lifted. Figure 6.6 presents a hypothetical gesture path for the intended word $\mathtt{slowly}$ that, as with the earlier touch typing example, does not quite hit the key targets for all letters of the word. As we have seen with auto-correction in the previous section, gesturing on a small mobile keyboard to a specific desired letter position can be somewhat imprecise, so the system must make its best guess about the intended input, and this can be informed by both

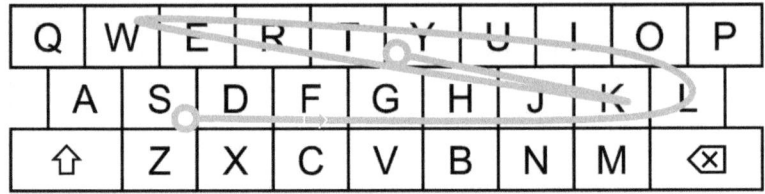

Fig. 6.6 Path traversed using gesture-based input on a virtual QWERTY keyboard for the target word `slowly`, starting at S and ending at Y. Source: own work.

how well the gesture corresponds to the ideal path for the word and how well the word is predicted from the context.

Since multiple letters may fall on the path associated with a word, some ambiguities naturally arise in this input method. For example, we noted in Chapter 5 that, according to some, the top row of the QWERTY keyboard (QWERTYUIOP) was at least partly chosen to make it easy for salespeople to demonstrate by typing the word *typewriter*, since all of the required letters fall on that single row. Gesture-based input of the word 'write' on a QWERTY keyboard starts at 'W', passes through 'R' en route to 'I', then returns through 'T' to 'E', all along that horizontal row. The same path also matches the word 'wire' (start at 'W', over to 'I' then back to 'E'). Beyond this, assuming that there might be some minor imprecision on the starting and stopping points, the gesture is very close to what is require for 'quite', 'quote', 'wow', 'woe' and a few other words that might appear in a standard English lexicon.

This can be treated similarly to touch typing as presented in the previous section: given the gesture, find the word with the highest conditional probability. Indeed, Ouyang et al. (2017) present their decoder as handling either touch or gesture input. Direct end-to-end modeling of this sequence-to-sequence mapping (input gesture, output text) faces the same obstacles as touch typing. For a noisy-channel approach, as presented in the previous section, what must change is the spatial model, i.e., estimating the probability of the input path given the candidate word.[18]

One major difference between gesture-based input and touch typing is the lack of a well-defined literal string in the former. Recall that the literal string in touch typing is the string formed from the letters where the touchpoint fell within the bounding box of their associated keys. Because gesture-based input traverses a range of keys without lifting the finger, there are many possible literal strings—see the list of words on the same path as `write` noted above. As pointed out in Ouyang et al. (2017), this makes decoding of gesture-based input more akin to speech recognition—which also lacks the notion of a literal—than touch typing. In any case, similar stochastic inference methods are applicable to this sort of input method as well.

---

[18] We refer the reader to that paper for details.

One key consideration for both gesture-based and touch typing input (not to mention speech recognition) is personalization, which we address again in Section 6.4.5. Before going there, however, we will examine the use of predictive models during composition for those for whom touch typing and/or gesture-based input are not accessible.

6.4.3 Co-construction and efficient gesturing in AAC

At the beginning of Chapter 5 we discussed a scene from the series Breaking Bad that involved a caregiver working with an individual (Hector) to produce a message using only a 6x6 physical letter board and a call bell. Hector's intended message was 'need dea', and the caregiver used row/column scanning over the letter board to determine each target letter. Scanning is a method for indirect selection, whereby an item is eventually chosen by answering a series of yes/no questions that narrow the alternatives until there is only one—see Section 5.5.2 where this method was introduced. The benefit of this sort of interaction is that it only requires indication of yes or no, such as with a call bell or another kind of (binary) switch. The Breaking Bad scene involved a human caregiver, but such interfaces are common in speech generating devices, thus allowing individuals to type using a binary switch without human assistance.

We present the same 6x6 grid of letters as shown in Figure 5.1 in Chapter 5 with a subset of the letters highlighted in Figure 6.7. We will discuss the significance of the shading/highlighting a bit later, but the grid does provide a handy reference for visualizing how the scanning interaction might work in a computer interface. The grid consists of 6 rows, and each row has 6 items. In row/column scanning, a letter is selected by first selecting the row, then selecting the letter in the selected row. The computer interface can prompt a selection of a row by highlighting it, starting at the top row. If the desired letter is in the highlighted area, then the user can select it—for example, by pushing a button. If the desired letter is not in the highlighted area, then the user will indicate a lack of selection by not pushing the button. After a pause to give the user sufficient time to push the button if they so desire, the highlighting advances to the next row. Once a row is selected, each item in the row is highlighted in turn, until one is selected.[19] If we signify a 'yes' (e,g. button press) with 1, and a 'no' (lack of button press) with 0, then we can assign each letter in the grid a binary code that specifies how it is selected in row/column scanninng, such as that for 'n' which is '001000001'.

One benefit of the grid as presented in Figure 6.7 is that each letter occurs in a fixed, easy-to-find location, so that figuring out if the target letter is within the highlighted group is relatively easy—just find the letter, and when

---

[19] How the interface behaves when nothing is selected is a design choice of the interface, something we will ignore for now.

A	b	c	d	1	2
E	f	g	h	3	4
I	j	k	l	m	n
O	p	q	r	s	t
U	v	w	x	y	z
5	6	7	8	9	0

Fig. 6.7 An AEIOU alphabet (or letter) board, as used in Breaking Bad, with a selection of letters highlighted. Source: own work.

it is highlighted, press the button. However, with a fixed grid in this order, likely letters in English such as 'n'—but also particularly 's' and 't'—have relatively long codes ('000100001' for 's' and '0001000001' for 't') which are time consuming to specify. In Chapter 5 we pointed out that reorganizing the grid so that frequent letters are placed near the upper left corner of the grid can reduce the expected length of these codes, since the most frequent letters would be significantly shorter. Such a fixed frequency ordered grid would not, however, be able to leverage the predictions of models about likely next letters based on the context. For example, in English, if the letter 'q' is entered, the probability that the next letter will be 'u' is much higher than if a letter other than 'q' had been entered, since English orthography generally requires a 'u' following 'q'. Dynamically reorganizing the grid after every letter makes locating the target letter in the grid challenging and effortful, hence is not typically done.

One method for leveraging predictive models during scanning of a fixed grid is known as Huffman scanning (Roark et al., 2013, 2015), since it relies on Huffman codes (Huffman, 1952) to produce optimally short binary codes for letters in context. The scanning method achieves this by breaking the assumption that highlighted regions are contiguous in the grid. Instead, it highlights a subset of letters with probabilities that sum up to as close to 0.5 as possible. This makes each yes/no question maximally informative, thus leading to a quicker selection (shorter binary codes). The approach of the user remains the same—simply find the desired letter in the grid and push the button whenever it is highlighted.

To illustrate how this would work, consider the first word of Hector's target message: 'need'. Figure 6.8 shows the probabilities of the first letter in a sentence, calculated from the British National Corpus (BNC Consortium,

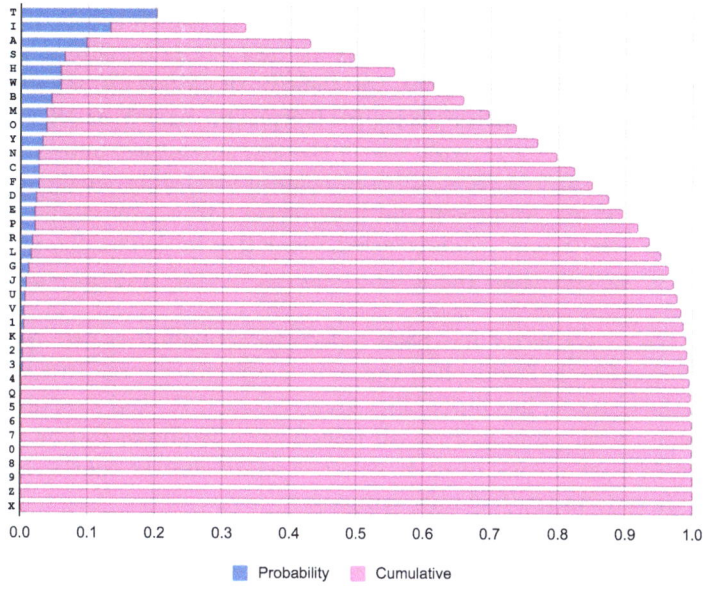

Fig. 6.8 Probabilities of first letter in a sentence, calculated from the British National Corpus.

2007), a curated, balanced corpus of 100 million words, in descending order of probability.[20] To calculate this, we extracted the first letter of every sentence, normalized for case, and removed everything except the 36 letters/digits in the grid in Figure 6.7. The graph presents both the probability of the letter in dark blue, along with the cumulative probability of that letter along with all the letters above it in light blue. From this chart, we can see that the four letters with the highest probability (T, I, A, and S) have a cumulative probability near 0.5, i.e., almost half of the sentences start with one of those four letters. If we add a couple of lower probability options (V and 5) to the set, we reach a collective probability of 0.49992, very close to the optimal 0.5 split. These are the six letters highlighted in Figure 6.7.

The target letter (n) is not in this set, meaning it is not highlighted, so, assuming an automatic scanning method (switch activation selects the

---

[20] Few people will produce English with a similar distribution to this corpus, but it suffices for this example. In the real world, having estimated probabilities close to what is expected from the user is central to such approaches, something that makes personalization such an important topic, something we touch on again at the end of the chapter.

highlighted subset), the user would not push the button.[21] Excluding these 6 letters/digits from our set, we can split the remaining 30 letters/digits in the same way: find a subset that accounts for half of the remaining probability mass. Using a similar procedure, we identify the following set of letters to highlight, based on the British National Corpus probabilities: {H, W, B, M, O, J, 6, 0}. The target letter is again not in the set, so another non-button-press. The next set {Y, N, C, F, U, K, Z} does contain the target letter, so a button press selects the set. And so on, until the target letter is selected. Given these probabilities, the binary code to select the letter 'n' is 001101, which has a length of 6, shorter than the 9 bits required in row/column scanning with this grid.

Once the target letter has been typed the probabilities are updated and scanning resumes. Following N at the start of sentences in the British National Corpus, the letter 'o' has probability greater than 0.72, and the (target) letter 'e' is second with probability 0.16. Hence, with Huffman coding, the target letter would have a code of 01 to select,[22] again shorter than even the very short 3 bit code in row/column scanning.

By combining a static grid and dynamic highlighting in the interface, searching for the target letter is easy (always in the same spot) and dynamic probabilities can be leveraged to speed up scanning. However, this is really only the tip of the iceberg in terms of what might be possible with effective predictive modeling, which becomes apparent in human-human co-construction scenarios. There is some attempted co-construction in the Breaking Bad scene presented at the start of Chapter 5, when the caregiver starts asking about possible completions for the second word ("Is it 'dear'? Is it 'deal'?"). Human-human co-construction involves the communication partner eliciting the intended language via yes/no questions, perhaps initially to determine letters via a scanning procedure, but not exclusively in that mode. Once a word, or even a phrase, seems likely to the communication partner, it can be presented to the individual for acceptance or rejection. If rejected, then the process can continue letter-by-letter.

One can think of this kind of co-construction interaction as a sort of 20 questions to build an utterance. It is the communication partner's task to choose questions that lead to the quickest production of the message, i.e., the fewest questions. Unless the context heavily constrains the likely utterance (such as the individual being themselves asked a yes/no question, hence likely responding with one of those two options), one will typically want to get the first letter or two before making a guess about the first word of a sentence. However, once some of the sentence has been produced, the intended content may become clear, so that predictions of larger units of text can be profitably made.

---

[21] See the papers on Huffman scanning (Roark et al., 2013, 2015) for methods that account for the probability of making and repairing errors, which we will ignore here for ease of exposition.

[22] We invite the motivated reader to work out why this would be the binary code.

A skilled communication partner, with knowledge of the individual, their tendencies and interests, as well as the context of communication, can make informed guesses about the intent, greatly speeding communication. Within a computer interface, this could be operationalized as word or phrase completion options being presented to the user when the probabilities of these completions are high enough to warrant it, provided the probabilities provided by the model are a good match to the user's behavior. With the advent of large language models and all their predictive power, one might expect a system making use of them to make useful predictions in such a context. A model may still be just as confused as Hector's caregiver when given the DEA token—it requires familiarity with that individual's history to clearly recognize the intent, something large language models would not provide right out of the box. We return to the personalization issue at the end of the chapter.

Before that, however, we will discuss some alternative text input interfaces built around predictive modeling. In this section, we advocated for a co-construction interface that alternates between letter, word and phrase predictions as elicitations for a yes/no response. Building such an interface that is intuitive and easy to use would be challenging. In the next section, we will discuss some systems with unusual and instructive text entry interfaces where predictive models are a primary component of the system.

6.4.4 Alternative predictive input interfaces

We begin our discussion of alternative predictive input interfaces with the *Dasher* system, which was first prototyped and documented in the late 1990s (Ward et al., 2000). Its inventors were part of an information theory group at Cambridge University led by David MacKay, and the system centered around *arithmetic coding* of strings of letters based on a model's probabilities. Notably the letters in such an approach are displayed in alphabetic order, so that they are easy to find; the size of the region associated with strings is allocated proportionally to their probability, making likely strings easier to select.

The Dasher interface places the letters along the right periphery of the screen, and typing occurs by navigating towards the regions associated with the target letters. Navigation can occur via joystick, mouse or even tracking of eye gaze. For example, in Figure 6.9(a), which shows the interface as typing begins, the letters from A-Z are listed in alphabetic order along the right periphery. The red line, which points in the direction of navigation, is pointing towards the letter T. Note that not every letter is explicitly displayed in the interface; in fact the least likely letters (including J, Q and Z) are omitted. Even so, since the letters are sorted alphabetically, one knows where the omitted letters should be, and navigating towards their neighbors

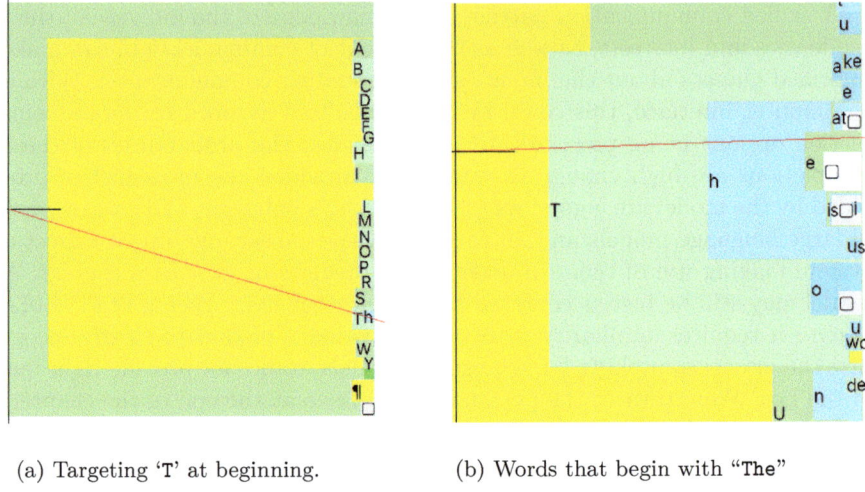

(a) Targeting 'T' at beginning.   (b) Words that begin with "The"

Fig. 6.9 Dasher interface (a) when typing is just beginning and (b) as typing progresses to select strings that start with "The". Source: screenshots of software window.

will eventually increase their probability to the point where they appear in the interface.

As navigation moves to the right, the regions where one is navigating grow, as though one were moving into those regions. Likely subsequent letters then begin to appear, with their regions also proportional to their probability. For example, in Figure 6.9(b), as the T region grows, inside of it we can see the next letter h as a likely option (hence with a large region inside of T), and after that e, which is where one would navigate to type any word beginning "The".

How does the arithmetic coding work? Suppose we had a line of length 1 (choose your favorite unit of measure), and we placed letters in alphabetic order into intervals along that line. Each letter goes into an interval proportional to its probability as the first letter in a sentence. Referring back to the British National Corpus probabilities presented in Figure 6.8, we see that the letter A, which comes first in our alphabet, has probability of about 0.1 of being first in a sentence in this particular corpus (0.096243 to be more precise). The letter A would thus be placed in the interval $[0, 0.096243)$, which starts at (and includes) 0, and ends at (but does not include) 0.096243. The next letter, B, is less than half as likely, with probability 0.045367. Its interval would begin at (and include) the point where the A interval stops, and end at (but not include) the beginning point plus 0.045367, namely: $[0.096243, 0.14161)$. Then the C interval would begin, again proportional to its probability and so on.

Once the intervals associated with the first letters have been established, then intervals for the second letters—inside of the intervals for first letters—

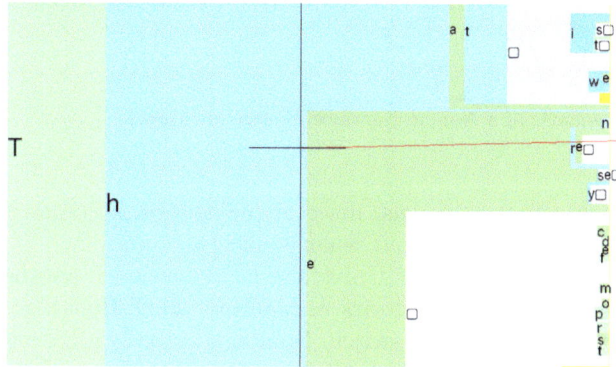

Fig. 6.10 Dasher interface when navigating to the word **There** followed by whitespace □. Source: screenshot of software window.

can be established. Given a first letter, say A, which has an established interval $[0, 0.096243)$, each letter following A would take a sub-interval inside this established interval, again proportional to the probability of that second letter following the first letter. For example, seeing a second A after the first A at the beginning of a sentence does not have a particularly high probability (just 0.00076 in the same corpus). In contrast B following that first A is more than 20 times more probable than that (0.01535). Since the established interval associated with the first letter A is of length 0.096243, the conditional probability is normalized to that length to establish the sub-intervals. Hence the interval associated with AA is $[0, 0.00007314)$ and the interval associated with AB is $[0.00007314, 0.00155)$. The same process continues as much as needed, so that arbitrary strings can be associated with (eventually very small) intervals.

Abstractly one can think of an intended sentence in this approach as a point along the right periphery of the screen, and typing involves navigating there. The colored regions allow one to stay oriented en route, and as one gets closer the string reveals itself. Figure 6.10 shows the path to type the word 'There' followed by whitespace, which is signified by □ in the interface. If one were to continue navigating to produce a whole sentence, then words that are likely to follow 'There' would be allocated high probability (hence large regions) and would thus be easier to navigate towards.

Typing fast in this interface is achieved by navigating quickly to the desired text, which is a bit like downhill skiing—one can get too far out over the skis and have trouble navigating to the original intended string as the letters whip past. In fact, one can get into autopilot mode with forward momentum, which will lead to some point that is associated with some non-intended string.[23] Finding the balance between speed and accuracy within the Dasher interface requires physical effort that may not be as accessible as, say, a button press.

---

[23] See the discussion of agency in Section 6.4.5.

Fig. 6.11 Schematic of a Nomon interface. Source: own work.

With better and better models, however, navigation becomes easier due to more accurate predictions about what is likely to come.

The same lab at Cambridge University later developed another text entry interface, called *Nomon* (Broderick & MacKay, 2009; Bonaker et al., 2023), intended for use with a single switch. Nomon is actually a general item selection mechanism, which can be applied to other selection tasks, but is easily specialized to text entry by associating the selectors with letters and words. Each item in the selection set is associated with a small clock on the screen to its left, with two arms on the clock: one always pointing straight up to noon, and the other rotating clockwise. For example, in the schematic presented in Figure 6.11, there are three items to select (A, B, C) and each has a clock with the rotating arm at a different position. All rotating clock arms rotate at the same rate, and to select an item, the user activates a switch (e.g., presses a button) when the rotating arm lines up with the noon arm in the target item's clock. Perfect precision is not required, and it may require multiple switch activations for the probability of selection to reach the required threshold, particularly if there are many items (and hence relatively small differences in timing for distinct items).

Given the flexibility of placement of the clocks, virtually any configuration of items can be accommodated, and the probabilistic mechanism that establishes the probability of items straightforwardly incorporates a language model. The above citations provide example interfaces that include placing likely words starting with a letter next to that letter, to make discovery of even word completion candidates easy. Since every possible item has the same sort of clock, rotating at the same rate, the effort required of the user by the interface really resides in finding the intended item and synchronizing the button press to the position of the rotating arm, possibly multiple times. As the quality of the model improves, as with earlier interfaces, the quality of predictions will yield a faster and easier interface.

We close out this section with perhaps the simplest single switch interface, which is *linear scanning*, i.e., presenting one item at a time with a single switch activation to select the presented item. Such an interface has a single location for the user to attend to, and items appear at that location for selection until one of them is selected. The items could be letters, words or even phrases. No visual search is required to find the location of the target or special regions of the interface for dynamic word predictions or completions— everything is presented in the same single location.

We will discuss two versions of this interface, the first requiring the user to press a button when they see their target item. This sort of serial visual presentation of items really only makes sense if the items are presented in descending order of probability, i.e., likely items first. Otherwise, the individual will often wait for a long time for their intended letter to be presented; and words or phrases have no natural way to be incorporated. Even with good predictive models, however, there will arise cases where unlikely letters and words need to be typed, and this method will require the user to wait a relatively long time to access those items. Having target items often appear quickly but sometimes very slowly may have good average performance but still be frustrating for the user.

Another version of such an interface uses rapid serial visual presentation (RSVP), at a rate that is too fast to allow enough time for a button press. Instead, intent can be detected via EEG signal processing from involuntary neural responses when the target item is presented (Orhan et al., 2012). This is a non-invasive brain computer interface (BCI) that, like Nomon, relies on predictive models to improve the system's ability to detect the intended letter. Because the items are presented so quickly—no voluntary response is required—many items can be presented in relatively short bursts, which addresses some of the issues with the approach when button presses are required. Even so, this sort of interface is primarily of utility for individuals for whom volitional switch activation is challenging or impossible. BCI interfaces, both invasive and non-invasive, do have the promise of extending accessibility to many who have few other options, and this remains an active area of research.

6.4.5 Considerations in the use of LLMs

We conclude the chapter with a short section covering some of the challenges and opportunities presented by large language models (LLMs) in this area. As this is being written, progress is being made very rapidly in terms of model capabilities, so there is some danger that this section will become out of date more quickly than other parts of the book. However, some of the issues involve a tension between the size of models and the typically local, low-latency processing required for assistance with text entry. While technical solutions will be found to specific impediments to use, this tension will likely persist for some time. Hence this is an appropriate topic to conclude a chapter on writing under space and time pressure.

The current paradigm of generative AI is to pre-train very large models which can then be specialized to perform desired tasks, either through *fine-tuning* the model with some supervised (even potentially synthesized) task-specific data or by including explicit instructions to the model. Pre-training the most capable models involves a scale of data and model that precludes casual or frequent updates—in other words, these are not models that individuals will be training and storing themselves. For example, the abbreviation expansion system of Cai et al. (2024) discussed in Section 6.3 uses a model with 64 billion parameters that was fine-tuned into two separate models to improve the efficiency of serving. As they note, the models require serving via an internet connection, which comes with a latency cost. If the overall text entry speed is relatively slow, such as in some AAC settings, the extra latency may not be an issue. However, generally mobile keyboard latency is expected to be sufficiently low so as to be imperceptible. Ouyang et al. (2017), whose work we cited when discussing auto-correction in Section 6.4.1, mention 20 milliseconds as the expected latency to produce visible output from a key press, and they use this as the rationale for having all processing occur on device. Longer latencies are perceived by users as a lag indicating a lack of system responsiveness. This suggests that, at least in the near term, the appropriate use of LLMs will be for conversion of compact input text rather than as a predictive model during composition of that text.

Conversion of input text to correct orthography or to another script is a task that may come with relatively strict constraints on what constitutes an acceptable result. For example, the expectation may be that the result contains the same words as the input text, simply normalized to a more conventional form. Deletion, insertion or significant change of one or more words may change the text in ways that are unwanted, yet this sort of creative revision is a relatively common feature of text elicited from LLMs. Significant work will be required to ensure that such systems follow instructions closely and reliably.

Relatedly, when text entry can rely on relatively little input to produce a relatively high volume of output, serious questions can arise regarding whether the resulting text originated from the individual. Examples might include using the Dasher interface discussed above without changing the orientation, i.e., just typing whatever text that appears as one navigates in a straight line at a particular orientation. Asking an LLM to write an essay on a topic is a relatively short input that can produce a lot of text, and few would argue that the resulting essay originated from the individual even though they requested it. Less egregious examples involving "good enough" paraphrasing of intended utterances may speed communication in ways that are largely beneficial, even if not phrased exactly as the individual would have, but it should remain within the individual's control to accept or reject content that has no direct correspondence in the input.

The issue of communicative agency has been around for as long as individuals have had writing assistance, and is not new to the use of predictive models. The low-tech practice of so-called *facilitated communication* provides a cautionary tale of how even well-intentioned assistance can lead to text that did not originate with the purported author. Briefly, facilitated communication was the practice of providing physical assistance to an individual's direct selection gesture, for those unable to complete such gestures on their own. The use of this method resulted in the production of surprisingly literate text from individuals who had been considered severely intellectually disabled.

By the mid-1990s, however, extensive controlled experimentation (see, e.g., Montee et al., 1995, and the references therein) demonstrated that the facilitators were responsible for the generated text rather than the individuals they were assisting. For example, when performing a picture identification task, if the facilitator and the individual were shown different pictures, the resulting text would identify the picture presented to the facilitator rather than the individual. The guidance exerted by facilitators was not typically conscious, rather the sort of unconscious movement that drives Ouija boards and similar phenomena. Similar issues can arise in partner- or computer-assisted co-construction when extensive prediction is relied upon.

Large language models encode a lot of general linguistic and world knowledge which are of high utility in making the kinds of predictions they are being asked to make. However, unless you are a public figure, they do not know much of anything about you as an individual, so any kind of personal skew of the predictions will need to come from elsewhere. For example, the model will not know if you are a member of the ACL special interest group on Speech and Language Processing for Assistive Technologies (SLPAT), so if you try to type "slpat" in a text message it may auto-correct it to *splat*. Incorporation of an individual's personal vocabulary and collocation patterns into the predictive models is known as personalization. Systems are constructed to build locally stored (and private) models from the text that is entered using the system, which can then be used to improve the system's predictions. The Dasher system (Ward et al., 2000) we discussed above had a particularly elegant character-based language modeling method to add newly typed text to their models; and the Ouyang et al. (2017) work we have mentioned also discusses the incorporation of dynamic local models into the text entry system. Combining such local models with off-device LLMs will likely require some clever engineering, particularly since such methods will have to occur on the device and under latency constraints.

Which brings us to the final consideration that we will mention which is privacy. We create text for a broad range of reasons and often share the results with select others—texts, emails, social media or blog posts, articles or books. Who we share it with—sometimes nobody—and under what conditions are decisions that are reserved for us to make ourselves. We typically make use of software to create the text—text editors, mobile keyboards, etc.—and that software is (hopefully) constructed to be privacy preserving, i.e., the text is only shared as dictated by the individual. The use of predictive models during composition, whether they are on the local device or accessed via an internet connection, must fall within that umbrella of trust and individual control.

The personalization methods mentioned above typically contain patterns of usage that reveal a lot about the content of the text that was produced. These patterns are absolutely useful for informing the predictions that can make text entry faster and more accurate, but they may also be used to potentially reconstruct specific text that was entered using the software, hence these personalization models must be treated with the same trust and control as the text that was produced. The same goes for data that is sent off-line to elicit predictions from an LLM. In order for the LLM to make its predictions, it needs to be provided with the context, which in this case means that to predict the probabilities of various possible next words or phrases, it needs to be given the context of what has already been typed.[24] If this sort of interaction forms part of the text entry system, then the information passed back and forth between the user and the system must fall under the same sort of privacy control.

Chapter 8 contains further discussion of the use of LLMs in writing.

---

[24] Also *prompting* methods may be used to improve LLM responses, which may include additional information beyond the current textual context.

Chapter 7
Diglossia and digraphia

7.1 Prelude

While it is difficult to come up with a solid estimate, and any estimate depends crucially on the definition of "bilingual", it is often claimed that the majority of the world's population is bilingual, i.e., speaks two or more languages in their daily lives (see, e.g., Ansaldo et al., 2008). These individuals will typically incorporate more than one language into their writing.

Many languages, such as Arabic, Chinese, and many languages of India, have local dialects that are distinct enough from each other to be mutually unintelligible. However, for these languages there is also a standard dialect used for writing and communication across these local dialects. Speakers in these communities often choose one language for specific social contexts and another for different situations, a phenomenon known as *diglossia* (Ferguson, 1959). For example, Arabic speakers might use Modern Standard Arabic (MSA) in official contexts, such as writing about politics or religion in their blogs, while they might use their native dialects, which often lack standard writing systems, when sharing everyday events with friends through text messages. In section 7.2, we will focus on cases of diglossia similar to Arabic, where one language has a writing system while the other does not.

Occasionally, multilingual speakers may mix several languages within a single text. For example, an email from Spanish-English bilingual speakers might contain content in both English and Spanish, sometimes within the same sentence, which is called *code switching*. For Spanish-English bilinguals, all such messages are typically written using the Latin script, allowing the same keyboard to be used for both languages. However, this is not generally the case for Chinese-English bilingual speakers, as the languages do not use the same script. Multilingual writers using language pairs that do not share the same script thus may face challenges in their writing process and develop different writing strategies compared to those whose languages share

B. Roark et al., *Tools of the Scribe*,
https://doi.org/10.1007/978-3-032-00831-2_7

the same script. In Section 7.5, we will explore multilingual writers who write in languages using different scripts.

Complicating matters, some languages are *digraphic*, i.e., they are written in more than one script. For example, Bosnia and Herzegovina has three official (and mutually intelligible) languages: Bosnian, Croatian and Serbian. While Croatian is typically only written in the Latin script, both Bosnian and Serbian are natively written in two scripts, the Cyrillic and Latin scripts. This is an instance of *synchronic digraphia*, where two or more scripts are used contemporaneously. Turkish is an example of *diachronic digraphia*, where the language was formerly written in a writing system using one script (Perso-Arabic) and now is written in a writing system using a different script (Latin). More recently, similar reforms—from Cyrillic to Latin—were implemented in former Soviet republics such as Azerbaijan. In Section 7.4, we will examine the case of Korean, where a society with two writing systems transitions into a single writing system society over time. Finally, in section 7.3, we will introduce India—home to 22 official languages and 13 scripts—illustrating how people write amidst extraordinary linguistic diversity.

In this chapter, we will explore writers who write in more than one language or more than one script. The choice of language or script in certain contexts or the switch from one to another within the same text is influenced by various technical, social, and individual factors.

7.2 Diglossia in writing

The term diglossia denotes the situation where speakers regularly communicate in two languages, with each language being used typically in well-defined circumstances. In principle the languages involved might be unrelated—Korean Americans communicating using Korean or English depending on the circumstance—or only distantly related—the comparable case for Hispanic Americans. However, the term is also often used to describe the use of two varieties of what is considered, sociolinguistically, to be the same language. Thus one may speak of Arabic diglossia , where people use MSA in formal situations, but revert to their native 'dialect'—which though genetically related to Standard Arabic, is often mutually unintelligible with it. In a similar vein one may speak of German diglossia (e.g. Standard German versus Bavarian, Swiss German, or Low German dialects); or Italian diglossia (standard Tuscan Italian versus, say, Neapolitan).

Of particular interest to this discussion, however, is *writing diglossia*, where writers who may be familiar with the standard orthography of the standard language, also want to write in their own dialect, for which no standard orthographic system exists. Such is the situation for, among many other cases, Arabic dialects or Swiss German.

In one sense the situation for dialect writers is no different from the situation faced by the earliest users of writing 5,000 years ago: A script exists, and some conventions exist for how to represent linguistic forms in that script, but there is no standardization to speak of, so the writer must rely on his or her own judgment on how to spell a given word. Thus in Sumerian, there were usually several ways one could spell the same word: 'beer' *kaš* could be written 𒁉 or 𒁉.[1]

Standardized orthography is something that develops over time. And insofar as any writing system is adapted to write a *particular* language variety, standard orthography also typically depends on one or more varieties becoming standard versions of the language. For example, German has been written in some form since the 8th Century—at that time Old (High) German, but it was not until the advent and spread of printing in the fifteenth and sixteenth centuries that a standardized spelling, based on East Central German dialects, started developing.

Standardized orthography, once developed, may come under the control of language academies, such as the the the Dutch Language Academy (Instituut voor de Nederlandse Taal), which has overseen numerous spelling reforms. One of these, in 1995 (Instituut voor Nederlandse Lexicologie, 1995; Neijt & Nunn, 1997), caused a fair amount of controversy as it proposed to upend many decades of convention in the spelling of certain morphological forms. Or standardized orthography may come about in a more organic fashion, with convention being arrived at by consensus: see Berg & Aronoff (2017) for a detailed account of the emergence of standardized spelling in English.

As noted above, the writer of a non-standard variety of a language is in many ways in the same situation as the earliest writers. A script (or perhaps several scripts) exists. Furthermore, conventions exist for its use for a standard variety of the language. The problem for the writer then is to decide how to apply those conventions, as far as possible, to their own dialect. This task can present multiple challenges. The dialect may have sounds that do not exist in the standard language, and one must decide how these are to be spelled. The dialect may have words that are not used in the standard language. This in itself may not be problematic if the standard orthography is fairly shallow, but if the writer is starting with a deep orthography, do they spell the dialect-specific word 'phonetically', or do they instead mimic what the standard orthography might do and attempt a deeper spelling? On the other hand, for words that do exist in both the standard language and the dialect, the writer may be tempted to adapt the standard spelling, even if it is not entirely appropriate for the variety used in the dialect, if only because the spelling is more familiar. A case in point is Taiwanese (Southern Min) *lâng* 'person', which is often written as 人 (Mandarin *rén*) 'person', since that character is very familiar: but etymologically the Taiwanese form is unrelated

---

[1] *The Pennsylvania Sumerian Dictionary* http://psd.museum.upenn.edu/nepsd-frame.html.

to Mandarin *rén*, but is instead the same etymon as *láng*, which is written
郎.

However one difference between writing in a non-standard language, and
writing in a language for which there is as yet no standard orthography,
is precisely the standardization itself. Simply put, standardization confers
a certain status to the standard system, so that in contrast dialectal forms
are felt to be inferior or corrupted versions of the standard. Historically this
has the situation exactly backwards: standard varieties of any language are
simply dialects—often an artificial compromise between dialects—that have
been raised in status. In most cases, other regional dialects or sociolects (such
as African American Vernacular English) are not derived historically from the
standard, and thus cannot be said to be 'corruptions' of it.

To take a specific example, for much of the early Middle Ages in Western Europe, literacy was equated with knowledge of and literacy in Latin.
Prior to the appearance of the first written texts in vernacular Romance
languages, starting with Old French in the late 9th Century, the regional
languages of the former Roman empire that eventually evolved into French,
Provençal, Spanish, Italian and so forth, were considered to be little more
than poor versions of Latin, to which they were obviously related, but from
which they obviously differed. Inevitably, regional variations in the spoken
language crept into written Latin. This led to the situation where in the late
8th century, Charlemagne, who considered these forms of Latin "intolerably
corrupt" (Price, 1971, p. 6), hired a team of expert consultants, among them
Alcuin of York, to help him rectify the situation and restore the Latin used in
the Church to its original "purer" classical form. But not surprisingly, this led
to the priests giving sermons that were unintelligible to most of their parish-
ioners, so that the Council of Tours, in 813, had no choice but to instruct
priests to provide translations into the local vernacular.

But again, Charlemagne had the situation backwards: Romance languages
were not "corruptions" of Classical Latin. Rather, Classical Latin was a
standardized variety of the language—Vulgar Latin—that people spoke as
a lingua franca in the Western Roman Empire. Romance languages evolved
from Vulgar Latin, not from the standard variety. Acquiring literacy during the Middle Ages required learning another language—a common situation throughout history: for example we saw earlier that Akkadian-speaking
Old Babylonian scribes needed, as part of their scribal training, to learn the
Sumerian language. When eventually written forms of the vernacular Romance languages appeared, writers developed their own spelling conventions
that were distinct from the conventions used to write Latin.

As noted above, German orthography was not standardized until quite
late, and so essentially all German texts of the Middle Ages were written ren-
ditions of regional varieties. Spelling systems for languages like Swiss German
are therefore simply a continuation of that tradition, albeit in a sociolinguistic
situation where a particular variety is now considered the standard. But while
using a non-standard form can be considered to be inferior, this sociological

valuation can also be turned on its head, so that use of a non-standard variety can be considered 'cool'. While there is no single standard for writing Swiss German, writing it in any form is a way of reinforcing Swiss identity (Ruoss & Schröter, 2020). This is exemplified by the publication of the 2012 Annual Report for the Swatch Group entirely in Swiss German (Swatch Group, 2012).

For Arabic, especially with the advent of social media, there has been a lot of interest in writing regional dialects. But writing colloquial Arabic, as opposed to the more refined Classical Arabic of the Qur'an, has a long history dating back to the very beginning of the Islamic period. Thus Holes (2018) notes that various mundane texts—"private and business letters, property deeds, marriage contracts, administrative surveys, lists and registers, passports, petitions, tax receipts, demands for payment, etc." (page 9)—have been preserved that are likely to be closer to the spoken form of the language. This is especially because such texts, having no literary or religious value, have not been 'normalized' over the centuries by countless editors, as has often been the case with literary or religious texts. Ferrando (2018) in the same volume discusses the spelling of a particular grammatical morpheme in Andalusi Arabic of the twelfth century, that has no counterpart in standard Arabic, and Kaye (1976) discusses pre-modern systems to spell Sudanese Arabic.

Recent studies, such as Khalil (2012) and Qudah et al. (2017), have found that Arabic speakers often use their local dialects when writing on computer-mediated communication (CMC) platforms, including short message service (SMS), emails, chat rooms, online forums, and social networking sites. When writing in their local dialects, Arabic speakers may use either the Arabic script or romanization with the Latin script. The idea of using the Latin script for Arabic dialects dates back to the 19th century (Yaghan, 2008). Since then, many proposals have been made. Although these proposals faced strong criticism from Arab nationalists and Islamic conservatives, the use of Latin letters for writing Arabic has become common and widely accepted. This practice, often referred to as "Arabizi" (a blend of "Arabic" and "English"), has become the main script for writing in CMC platforms.

The Arabizi system varies across Arabic-speaking countries, depending on the local dialect, but some general rules have been summarized by researchers such as Yaghan (2008) and Palfreyman & Khalil (2003). Generally, words are spelled based on their sounds rather than through a direct one-to-one mapping between Arabic and Latin characters. For Arabic consonants that have an equivalent in English, the corresponding English letter is used. In contrast, unique Arabic consonants are represented by numerals such as '2', '9', '6', and '3'. For example, the numeral '2' is used for glottal stop /?/). Although these digits are chosen for their visual similarity to the original Arabic characters, identifying the intended word can be challenging without familiarity with the spelling conventions. Short vowels, usually omitted in MSA, are sometimes represented in Arabizi to resolve ambiguity in meaning, but they may also be omitted due to character-length constraints.

Technical issues were among the most important reason for using Arabizi on CMC platforms. A survey conducted by Palfreyman & Khalil (2003), involving 79 university students in Dubai, found that 55% of students used Arabizi for the ease of typing on an English keyboard, while 30% used it due to difficulties accessing Arabic keyboards on mobile phones. Although most mobile devices now include Arabic keyboards, making the latter issue less relevant, Arabizi remains a popular writing system in CMC (Alsulami, 2019; Alghamdi, 2018)

How then did writers learn Arabizi spelling conventions? The same survey reported that the majority (70%) learned it from relatives or online acquaintances, while none of them learned it from websites or print materials. Without formal education or the announcement of a standardized orthography, variations in spelling are common. However, messages seem to be generally understood, as demonstrated by various posts on a website dedicated to Arabic teaching and resource sharing (Lingualism, 2020).

As mentioned in Chapter 6, romanization is also common in South Asia, at least in part for the same reasons as just outlined for Arabizi. However, the high levels of regional script diversity—in India in particular—leads to many overlapping multilingual and multiscript scenarios.

7.3 Script diversity in India

There are many languages spoken in India—22 'scheduled' languages with official status according to the government,[2] and many more with significant numbers of native speakers beyond those 22. English also has status as one of two (alongside Hindi) official languages of India. Additionally, however, India has a relatively high level of script diversity among (and even within) those languages. Thirteen different scripts are used in India for the 22 scheduled languages plus English, which is the sole language written natively in the Latin script. Hindi and 9 others[3] are written in the Devanagari script, though two of those (Kashmiri and Sindhi) also are written in the Perso-Arabic script. Urdu, which, like Hindi, is a register of Hindustani (hence mutually intelligible with Hindi), is written in the Perso-Arabic script. The Bengali script—like Devanagari, a Brahmic script—is the native script of two more scheduled languages, Bengali and Assamese. Six of the other languages have their own Brahmic script: Gujarati, Kannada, Malayalam, Odia, Tamil and Telugu. The Brahmic scripts Meitei and Gurmukhi are used for Manipuri[4] and Punjabi, respectively. Finally, Santali is written with Ol Chiki, an alphabet created in modern times (1925) specifically for Santali. Many more languages that are

---

[2] https://www.mha.gov.in/sites/default/files/Eighth_Schedule.pdf

[3] Bodo, Dogri, Kashmiri, Konkani, Maithili, Marathi, Nepali, Sanskrit and Sindhi.

[4] Manipuri is also commonly written in the Bengali script.

not on the schedule use Devanagari or, to a lesser extent, Bengali scripts, and across national boundaries, Punjabi is also written natively in Shahmukhi, a Perso-Arabic script.

Some of the digraphia arises in this complex patchwork due to political boundaries, such as Punjabi being written in Gurmukhi in India and Shahmukhi in Pakistan. Of course, Punjabi is also (informally) written in the Latin script in both India and Pakistan, so Punjabi speakers who do not share the same native script can likely communicate via romanization. Even inside of India, state boundaries can cause this sort of digraphia among a language community. For example, Konkani is the official language of the smallest Indian state, Goa, where the official script of the language is Devanagari. Konkani, however, is spoken in a wider region than Goa, including in the much larger neighboring state of Karnataka, where Kannada is the official language. As a result, Konkani speakers in Karnataka write in the Kannada script rather than Devanagari.

As a legacy from the colonial period, English remains an official language of India and is a lingua franca in the region. Brandt (2020) argues that the Latin script is the *scripta franca* of South Asia, largely due to the fact that it can be read everywhere. This de facto status for the script stands despite historical resistance to adoption of the Latin script as a standard in India, which has been repeatedly proposed (Brandt, 2020). But the lack of a standard orthography in the Latin script for Indian languages leads to the same sort of informal and varied usage associated with dialectal writing in diglossia presented in the last section. The prevalence of Latin script usage for all of the languages listed above means that, even those languages that have a single long-standing native script are fundamentally digraphic.

7.4 Journey from digraphia to a single writing system

Before the creation of the Korean alphabet, known as Hangul, Korean primarily relied on Chinese characters (Hanja) for written records. Even after Hangul was introduced in the 15th century in the Joseon Dynasty, both Chinese characters and Hangul were used in Korea. It was not until the 20th century that Hangul became the exclusive script. During this time, the ruling class men mainly used Chinese characters, while Hangul was associated with women and the lower classes. Thus pre-modern Korea displayed characteristics of digraphia where two scripts were used by a single language community under different conditions, with Chinese characters having more prestige (DeFrancis, 1984).

This phenomenon appears similar to the diglossia discussed in Section 7.2. However, upon closer examination, it differs in two major ways. First, premodern Korea was a monolingual society, using only the Chinese characters for writing, with few people actually speaking the Chinese language.

Secondly, pre-modern Korea used two distinct scripts for writing, whereas diglossic societies typically used one writing system for the standard language, with the native dialect borrowing writing from the standard language. Korean digraphia is more similar to the situation with Punjabi, where both Gurmukhi, a Brahmic script, and Shahmukhi, a Perso-Arabic script, are used to write the language. For the most part, however, these two scripts are not used within the same communities—Shahmukhi is used by Muslims and in Pakistan, while Gurmukhi is otherwise used in India. In contrast, Korean digraphia did not occur across such religious or political boundaries. In this section, we will focus on explaining the process by which a society with two writing systems transitioned to a single writing system based on the analysis of script usage in texts from the 15th to the 20th century.

Shortly after the 15th century when Hangul was introduced to the public, the government published several books to promote it. However, none of these books was written exclusively in Hangul; instead, they used mixed scripts, incorporating both Chinese characters and Hangul. We will explain the commonly used coding styles using a simple Korean sentence 봄에는 산에 꽃이 핀다, meaning "Flowers bloom in the mountain in spring". For clarity and ease of explanation, we used modern Korean spelling, spacing, and left-to-right writing style[5]. Table 7.1 presents this sentence in a variety of coding styles, which we will refer to when explaining key differences between styles.

Styles (a), (b), and (c) in Table 7.1 are written only in one script; (a) and (c) are written in Hangul script, while (b) is written in Chinese characters. (a) is the Korean vernacular form and is acceptable as a modern Korean sentence. In (a), all words including the Sino-Korean word 산 are written in Hangul. (b) is Literary Chinese, and it does not include any function words. In addition to the absence of function words, Literary Chinese uses a different word order although it is not demonstrated in this example. Most official documents in the Joseon dynasty were written in style (b).

Styles (d), (e), (f), (g), and (h) are all written in Chinese Character-Hangul mixed script. In style (d), each word is represented in either Hangul or Chinese characters, depending on its origin. The Sino-Korean word 山 is written exclusively in Chinese characters, while native words, including function words, adverbs, verbs and some nouns, are written in Hangul. (e) and (f) are similar to style (d), but they present all Sino-Korean words in both Chinese characters and Hangul. The distinction between (e) and (f) lies in the order of Hangul and Chinese characters. The order—whether Chinese characters come first or Hangul—is chosen based on the intended audience's familiarity

---

[5] Prior to the end of the Joseon Dynasty, Hangul was traditionally written vertically, from top to bottom. In the 19th century, a left-to-right writing style was introduced, eventually becoming the standard by the 20th century. The political climate and technical factors influencing these vertical and horizontal styles are discussed in Chapters 3.9 and 3.7.1. In this section, we used the left-to-right style while explaining the various coding styles to make explanations easier, whereas images of Joseon Dynasty texts were written in the top-to-bottom style.

(a) Korean vernacular sentence:

봄에는	산에	꽃이	핀다
⟨bom-e-neun⟩	⟨san-e⟩	⟨kkoch-i⟩	⟨pin-da⟩
Spring	mountain	flower	bloom

(b) Literary Chinese:

春	山	花	開
Spring	mountain	flower	open

(c) Korean pronunciation of (b):
춘산화개 ⟨chun-san-hwa-gae⟩

(d) Korean mixed script (1).
Chinese characters for Sino-Korean words, with Hangul for all native Korean words:
봄에는 山에 꽃이 핀다

(e) Korean mixed script (2).
Same as (d), with each Chinese character followed by its Hangul pronunciation:
봄에는 山산에 꽃이 핀다

(f) Korean mixed script (3).
Same as (d), with each Chinese character preceded by its Hangul pronunciation:
봄에는 산山에 꽃이 핀다

(g) Korean mixed script (4).
Literary Chinese (b), accompanied by a parallel line of Hangul pronunciation of the Chinese characters:
春山花開
춘산화개

(h) Korean mixed script (5).
Literary Chinese (b), accompanied by a parallel line of Hangul translation, instead of its pronunciation:
春山花開
봄에는 산에 꽃이 핀다

Table 7.1 Coding styles of the Korean sentence "봄에는 산에 꽃이 핀다".

with Chinese characters. In styles (g) and (h), a complete Literary Chinese sentence is initially presented, followed by Hangul to provide its pronunciation or a full translation in the Korean vernacular form. Although both styles offer two complete representations of a sentence (all Chinese characters and all Hangul), style (h) is much easier for Hangul readers to understand. Readers who are not familiar with Chinese characters cannot understand the meaning in style (g), since the sentence is written using Chinese words and word order.

Figure 7.1 displays an image of Seokbosangjeol, which is the first Buddhist text written in Hangul, published in the 15th century. Parts written in a combination of Chinese characters and Hangul are highlighted with red boxes, while Hangul characters are specifically marked with blue circles. This text was written top to bottom using coding style (e), in which Chinese characters are followed by Hangul pronunciation.

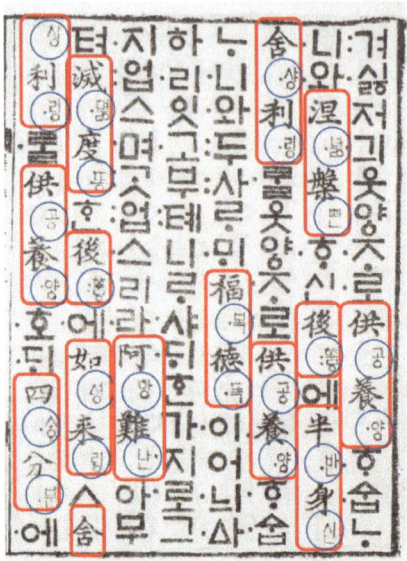

Fig. 7.1 Seokbosangjeol written in style (e). Source: Kyujanggak Institute for Korean Studies, Seoul National University. Image is in the public domain.

These different styles of mixed script have varying impacts on the interpretability and readability of the text. Chung (2009) evaluated the difficulty of interpretability for readers who only know Hangul. When these styles are ranked from most difficult to easiest, the proposed order is (d) > (g) > [(e) and (f)] > (h). Both (d) and (g) are challenging for Hangul-only readers to grasp the meaning, but style (g) at least allows them to recite the sentence. Therefore, style (d) is suited for readers who are already familiar with Chinese characters, while style (g) is intended for those studying Chinese characters. Style (h) is the easiest because even readers who do not know any Chinese characters can understand the meaning.

Soon after the promulgation of Hangul, it was used in various types of books, such as books designed to introduce the script, as well as books containing content that the government wanted to convey to the general populace. In particular, it was used in books explaining the phonetic values and usage of Hangul[6], as well as in Chinese character education books[7], religious

---

[6] E.g., *Hunmin Jeong'eum* (Proper Sounds for the Education of the People).

[7] E.g., *Dongguk Jeongun*, which explains the pronunciation of Chinese characters.

books[8], technical books[9] for commoners, and other pedagogical works[10] for women and commoners.

As previously mentioned, there were numerous styles of mixed script, each affecting readability differently. Consequently, for each book, a particular style was selected based on the target audience and its goals. For example, style (g) was used for content that was needed to reach the widest audience, including the lower classes. This included topics that were particularly instructive, such as the grace accorded to parents, the virtues people should uphold, and the penalties for violating these virtues.

Unlike books or official documents, personal letters were often written in style (a), using only Hangul. When one of the parties involved in the correspondence was not an adult male from the ruling class, the letters were often written solely in Hangul. Figure 7.2 shows a letter from a ruling class man to his young nephew. All the sentences were written entirely in Hangul. This serves as a good example to demonstrate that everyday content could be communicated entirely in Hangul, without Chinese characters, even when the communication involved members of the male ruling class.

In all the mixed script styles from the examples in Table 7.1, except for (d), Sino-Korean words appear twice: once in Hangul and once in Chinese characters. This duplicated representation of the same information of course led to inefficiencies in writing. The redundancy in styles (g) and (h) is even more pronounced because each sentence is presented in its entirety in both scripts. While this issue is less stark in our example because it contains only one Sino-Korean word, Korean texts typically include a much higher proportion of these words, leading to significantly greater inefficiency in average texts, and much duplication of information.

Quantitative analysis based on a modern Korean text corpus indicates that Sino-Korean words constitute approximately 66% of the words (Heo, 2010). While this proportion is calculated based on modern Korean texts, considering the frequent use of Sino-Korean words during the Joseon Dynasty, texts from that period would have included a comparable or even higher proportions of Sino-Korean words. Consequently, if all Sino-Korean words were written in both Hangul and Chinese characters, the length of the text and the effort required to write it would increase significantly. For ease of understanding, Figure 7.3 provides part of the Constitution of the Republic of Korea in both style (a) and (f)[11]. The part written in style (a) comprises

---

[8] E.g., Translations of Buddhist texts to make religious teachings more accessible to the general populace.

[9] Books disseminating practical information such as medical knowledge or farming technology.

[10] E.g., books designed to teach Chinese characters easily to children or books designed to teach Confucian doctrine to women from the ruling class and common populace.

[11] As shown in Figure 7.1, the sizes and positions of Chinese characters and Hangul differ in the original style (f). However, since the main purpose is to compare the text lengths between style (a) and style (f), Chinese characters are placed in parentheses, which is

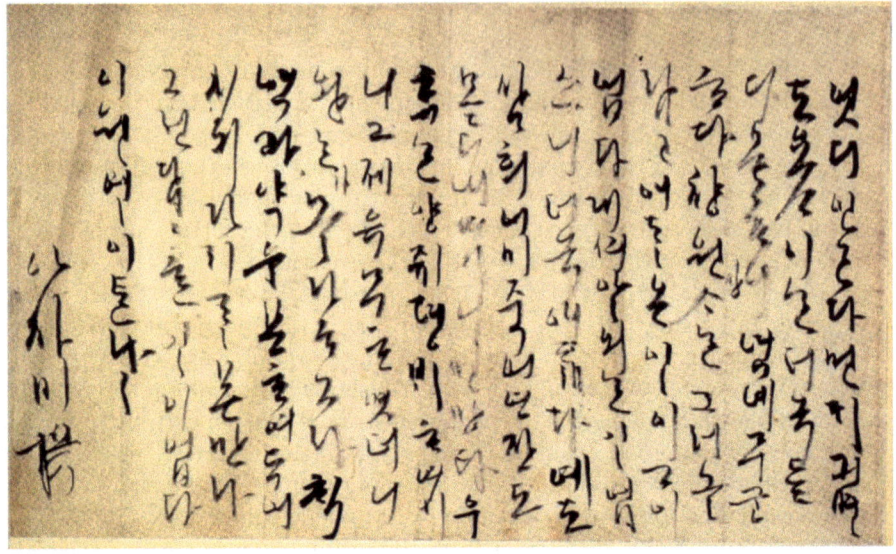

Fig. 7.2 A letter written by a Joseon Dynasty scholar (1595–1682). Source: Jangseogak Archives, The Academy of Korean Studies. Image is in the public domain.

341 syllables, while the same part in style (f) comprises 513 syllables. Style (f) is approximately 1.5 times longer than style (a).

From the 17th to the early 19th centuries, Hangul was used in numerous texts. Its use expanded not only in the number of volumes but also to encompass new genres. According to Han (2015a), these genres included translations of classical novels, travelogues, biographies, historical records (such as those documenting wars, emergencies, or diplomatic missions), diaries, and cookbooks. However, there was a significant difference from the earlier period; most texts from this period were composed solely in Hangul, rather than using the mixed script. This shift led Korean society toward more prototypical digraphia, where the ruling class typically used texts written only in Chinese characters, while other classes used texts written solely in Hangul (Jeong, 2008).

The turbulent political climate of the 19th and early 20th century, particularly the growing influence of Japan and Russia, changed the perception of Hangul even among the ruling class. During the period when Korea was gradually losing its national independence, many members of the ruling class started actively using Hangul, aligning it with patriotic sentiments. Accompanying this trend, the Korean government officially promulgated the exclusive

---

more typical in modern texts (e.g. on Wikipedia pages) where Chinese characters are included alongside Hangul spellings.

- **Text in style (a)**

유구한 역사와 전통에 빛나는 우리 대한국민은 3·1 운동으로 건립된 대한민국 임시 정부의 법통과
불의에 항거한 4·19 민주 이념을 계승하고, 조국의 민주 개혁과 평화적 통일의 사명에 입각하여
정의·인도와 동포애로써 민족의 단결을 공고히 하고, 모든 사회적 폐습과 불의를 타파하며, 자율과
조화를 바탕으로 자유 민주적 기본 질서를 더욱 확고히 하여 정치·경제·사회·문화의 모든 영역에 있어서
각인의 기회를 균등히 하고, 능력을 최고도로 발휘하게 하며, 자유와 권리에 따르는 책임과 의무를
완수하게 하여, 안으로는 국민 생활의 균등한 향상을 기하고 밖으로는 항구적인 세계 평화와 인류 공영에
이바지함으로써 우리들과 우리들의 자손의 안전과 자유와 행복을 영원히 확보할 것을 다짐하면서
1948년 7월 12일에 제정되고 8차에 걸쳐 개정된 헌법을 이제 국회의 의결을 거쳐 국민 투표에 의하여
개정한다.

- **Text in style (f)**

유구(悠久)한 역사(歷史)와 전통(傳統)에 빛나는 우리 대한민국(大韓國民)은 3·1 운동(運動)으로 건립(建
立)된 대한민국(大韓民國) 임시(臨時) 정부(政府)의 법통(法統)과 불의(不義)에 항거(抗拒)한 4·19
민주이념(民主理念)을 계승(繼承)하고, 조국(祖國)의 민주(民主) 개혁(改革)과 평화적(平和的) 통일(統
一)의 사명(使命)에 입각(立脚)하여 정의(正義)·인도(人道)와 동포애(同胞愛)로써 민족(民族)의 단결(團
結)을 공고(鞏固)히 하고, 모든 사회적(社會的) 폐습(弊習)과 불의(不義)를 타파(打破)하며, 자율(自律)과
조화(調和)를 바탕으로 자유(自由) 민주적(民主的) 기본(基本) 질서(秩序)를 더욱 확고(確固)히 하여
정치(政治)·경제(經濟)·사회(社會)·문화(文化)의 모든 영역(領域)에 있어서 각인(各人)의 기회(機會)를
균등(均等)히 하고, 능력(能力)을 최고도(最高度)로 발휘(發揮)하게 하며, 자유(自由)와 권리(權利)에
따르는 책임(責任)과 의무(義務)를 완수(完遂)하게 하여, 안으로는 국민(國民) 생활(生活)의 균등(均等)한
향상(向上)을 기(基)하고 밖으로는 항구적(恒久的)인 세계(世界) 평화(平和)와 인류(人類) 공영(共榮)에
이바지함으로써 우리들과 우리들의 자손(子孫)의 안전(安全)과 자유(自由)와 행복(幸福)을 영원(永遠)히
확보(確保)할 것을 다짐하면서 1948년(年) 7월(月) 12일(日)에 제정(制定)되고 8차(次)_에 걸쳐 개정(改
正)된 헌법(憲法)을 이제 국회(國會)의 의결(議決)을 거쳐 국민투표(國民投票)에 의(依)하여 개정(改
正)한다.

Fig. 7.3 Preface of the Constitution of the Republic of Korea

use of Hangul in 1894. The government declared the principle that "laws and orders shall be based on the national script, with translations in Literary Chinese attached, or occasionally using a mixed script." This marked the beginning of the official policy for the exclusive use of Hangul, which came to be used in all areas. By this time, Hangul had become widespread, appearing in newspapers, magazines, academic books, and legal and administrative documents—areas where it had previously been rarely used or not used at all (Han, 2015b).

However, it was challenging to establish the exclusive use of Hangul. One of the major challenges was the large number of homonyms in Sino-Korean words. In Hangul-only text, all Sino-Korean words (e.g., 山 in [d]) are written in Hangul (e.g., 산 in [a]). This increases the ambiguity of the meaning because many Sino-Korean words have multiple homophones. For instance, the Standard Korean Language Dictionary (National Institute of Korean Language, n.d.) lists 31 different meanings for the term 사장 ⟨sa-jang⟩. After excluding meanings used in specialized areas such as Buddhism and law, the following 12 meanings remain:

(1) 死藏: Letting things go to waste without utilizing them where needed.
(2) 砂場: A wide and large sandy plain by a river or seaside.
(3) 社長: The chief executive officer responsible for company operations, holding the authority of the company representative.
(4) 社章: A commemorative badge agreed upon when forming an association.
(5) 社葬: A funeral conducted by a company.
(6) 查丈: A term used between the parents of two families in a marriage to refer to the counterpart who is of the same generational rank.
(7) 射場: A place equipped with facilities for archery.
(8) 師匠: A person skilled in academics or arts who is worthy of being a teacher to others, or someone who teaches academics or arts.
(9) 師長: A teacher and an elder.
(10)紗帳: A curtain made of thin and light silk.
(11)詞場: The society of literary figures.
(12)辭狀: A document stating the intention to resign from a position.

Each of these meanings is associated with different Chinese characters, allowing for easy disambiguation when the word is written in Chinese characters. However, if it is written only in Hangul, readers must rely on the context to determine the correct meaning. Furthermore, Sino-Korean words are more prevalent in fields like business, law, administration, and academia, and texts in these areas, when written only in Hangul, present greater interpretation challenges due to ambiguity. After removing Chinese characters, the texts become much harder to read.

Due to these challenges, many texts continued to be written in mixed script, leading to the coexistence of both Hangul-only and mixed-script texts. This coexistence was particularly evident in newspapers, where each agency adopted different styles to appeal to their target audience. Furthermore, some newspapers, like "The Korea Daily News", produced articles in both styles, selecting the style of each article based on the topic and potential audience.

It is important to note that the characteristics of mixed script during the early period of the move towards the wider adoption of Hangul varied significantly depending on factors such as the authors and genres involved. Texts in certain genres or written by certain authors were more similar to the modern Korean vernacular, while others were not. The differences were not only in the proportion of Sino-Korean words but also in grammatical structures. Han (2015b) categorized the mixed script styles based on the extent to which grammatical structures were used, as follows. Here, we base our examples on the same sentence about flowers blooming in the mountain in spring as we used in Table 7.1:

(b1) Literary Chinese sentence as it is, supplemented with Korean function words:
春山花開 하다

Word	1920s	1930s	1940s	1950s
we	0.3%	0.6%	0.0%	0.0%
ask	53.2%	11.7%	0.0%	0.0%

Table 7.2 The percentage of incorrect Literary Chinese usages in Korean newspaper 'Donga Ilbo'.

(d1) Korean mixed script.
Influence of Literary Chinese grammar on noun phrases:
春에는 山에 花가 핀다

(d2) Korean mixed script.
Influence of Literary Chinese grammar on verb phrases:
봄에는 山 에 꽃이 開하다

Style [b1] consists of Literary Chinese sentences that incorporate the Korean function word 하다 ⟨ha-da⟩ ('to do') to indicate citation. It is important to note that this style is not considered a form of Korean vernacular.

Styles [d1] and [d2] are variations of style [d] and considered a Korean vernacular form. However, both styles include Literary Chinese words that are not legitimate Sino-Korean words. For instance, in style [d1], native Korean nouns such as 봄 and 꽃 are replaced with Literary Chinese words 春 and 花 that are not part of legitimate Korean vocabulary. These conversions are awkward, resembling inaccurate translations between Korean and Chinese. This results in sentences that lack natural fluency in either language.

This translation-like mixed script emerged during the transition period and eventually disappeared as modern mixed script conventions became established. Han (2017) demonstrated that these changes occurred gradually from the late 19th century to the early 20th century, based on the analysis of 13,000 newspaper editorial articles. He selected the two common words 'we' and 'ask' and compared the distribution of native Korean words with their Literary Chinese counterparts. Similar to styles [d1] and [d2], the Literary Chinese versions were not valid Sino-Korean words, and using them made Korean sentences sound unnatural. Table 7.2 presents the percentage Literary Chinese usage. This percentage is calculated by dividing the frequency of Literary Chinese words by the total frequency of both Literary Chinese and native Korean words.

The use of these Literary Chinese words decreased from 1920 to 1930, and they completely disappeared after the 1940s. However, the distribution patterns of the two words were quite different. For the pronoun 'we,' the native Korean was already used exclusively from the 1920s onward. In contrast, for the verb 'ask', its Literary Chinese counterpart was more frequently used than the native word in the 1920s, but there was a substantial decline in its use during the 1930s.

After independence from Japan, both South Korea and North Korea implemented policies to promote the exclusive use of Hangul. In North Korea, the transition to using exclusive Hangul progressed relatively quickly, with the use of Chinese characters being abolished by 1949 (Lee, 2014). In contrast, the transition to the exclusive use of Hangul in South Korea took place over a relatively longer period. The government of South Korea enacted the 'Act on the Exclusive Use of Hangul' in 1948. It was similar to the principle established in 1897, which stated that official documents should be written in Hangul, although Chinese characters could still be added alongside. Even after this act was promulgated, mixed script continued to be frequently used in various texts. However, as younger generations became less familiar with Chinese characters, the use of texts incorporating them gradually declined. The widespread distribution of computers in the 1990s further accelerated the exclusive use of Hangul.

Why was exclusive Hangul adopted more quickly in North Korea than in South Korea? At the time of independence, there were no significant linguistic differences between South and North Korea, suggesting that South Korea did not face greater linguistic challenges than North Korea. The rapid transition in North Korea was due to the government's stronger commitment and investments. They not only abolished the use of Chinese characters but also implemented various policy measures to reduce semantic confusion, such as replacing many Sino-Korean words with newly developed native Korean vocabulary (Lee, 1982).

Today, although there are still texts that use mixed script depending on the audience or genre, the digraphia perspective that viewed Chinese characters as a superior script and Hangul as inferior has disappeared. Korea has transitioned into a linguistic community that predominantly uses a single script .

7.5 Code switching in multilingual writers

Code-switching is a phenomenon where multilingual speakers alternate between two or more languages or dialects within a conversation, sentence, or discourse. Speakers proficient in multiple languages switch between them based on context, audience, or conversational needs. Since the term "code-switching" was first introduced by Vogt (1954) in the field of linguistics, the factors that influence code-switching patterns and its functions have been widely studied in speech.

While much of the research has focused on spoken language, the exploration of code-switching in writing has remained relatively underexplored. However, with the rise of computer-mediated communication (CMC), this area is gaining more attention. As CMC has become increasingly popular and significantly impacted how we communicate, it has prompted a growing interest in understanding code-switching patterns within these digital texts.

In this section, we will explore code-switching patterns among languages that share the same scripts and those use different scripts. Before diving into the detailed discussion about code switching in CMC, we will start by summarizing types of code-switching. Poplack (1980)'s typology, which categorizes code-switching into inter-sentential, intra-sentential, and tag question based on the syntactic characteristics, has been widely accepted in the field. Intra-sentential code-switching occurs within a phrase or clause, while inter-sentential code-switching takes place between sentences. The tag question type involves incorporating a question tag from one language into an utterance in another language.

First let us consider code-switching patterns in languages sharing the same script. Barasa (2016) analyzed code-switching patterns in CMC texts from Kenyan university students. Kenya is a highly multilingual country with English and Swahili as official languages, along with more than 40 vernacular languages. Despite this linguistic diversity, these languages share the Latin script for writing, although standardized spelling systems for vernacular languages have not yet been established. Based on the analysis of the various genres of CMC texts, she found that code-switching occurred frequently, with approximately 59% of texts containing instances of code-switching. Bilingual code-switching was the most common, but trilingual code-switching was also significant, occurring in 37% of cases.

The following example, taken from Barasa (2016), is an instant chat between two students. For each word, we have indicated the language to which it belongs. English, Swahili, and Sheng were used during the conversation[12].

Student 1:	asa msupuu,	you there?
(Language used)	Sheng	English
(Literal Translation)	Hi pretty,	you there?

Student 2:	niko	poa	kabisa,	how are you doing?
(Language used)	Swahili	Sheng	Swahili	English
(Literal Translation)	I am fine completely,			how are you doing?

In these examples, inter-sentential code-switching occurred frequently. In the sentence from student 2, code-switching occurred three times, with each of the first three words being in a different language from the preceding word. A broader analysis of the texts showed that code-switching happened freely, without fixed patterns or orders. All texts were written in the Latin script, with no markers to indicate the language switch. Because the three languages use the same script, writers do not need to switch keyboards for each instance of code-switching. However, this places more effort on readers, who must identify the language of each word to decode the message.

---

[12] Sheng is a Swahili- and English-based slang rather than a separate language, but we will follow Barasa (2016)'s convention and annotate it separately.

Building on this understanding of code-switching within a single script, recent studies have explored code-switching involving two different scripts in CMC. For instance, Goh & Wu (2024) and Nakamura (2020) investigated Japanese-English code-switching, Lin (2005) and Gonzales & Tsang (2023) examined Chinese-English code-switching, and Srivastava et al. (2020) studied Hindi-English code-switching. Goh & Wu (2024), Nakamura (2020), and Gonzales & Tsang (2023) gathered CMC texts from bilingual participants and conducted linguistic analyses supplemented by interviews. Srivastava et al. (2020) examined one million Hindi-English bilingual tweets without surveys or interviews. Finally, Lin (2005) employed a survey-based approach. These studies revealed four different patterns of script switching:

(a) Mixed scripts:
  1. i think im going home tuesday and
     火曜日までプロジェクトで忙しいかも (Goh & Wu, 2024)
     (Translation) I think I am going to home Tuesday and I am busy with a project until Tuesday.
  2. 上個 project end up 成點 (Gonzales & Tsang, 2023)
     (Translation) How did the last project end up?
  3. 我今個 sem 唔計 lab 有兩日 有早堂 (Gonzales & Tsang, 2023)
     (Translation) I have morning classes for 2 weekday this semester, excluding laboratory sections.
  4. वो बाबा ढोंगी नहीं थे so better watch your mouth before blabbering (Srivastava et al., 2020)
     (Translation) That Baba (Spiritual Teacher) was not an imposter, so better watch your mouth before blabbering.

(b) English and romanization of non-English language:
  1. Lunch was not impressive- **yasai itame teishoku.** Mostly **moyashi**. (Nakamura, 2020)
     (Translation) Lunch was not impressive- stir-fry veggie set. It's mostly bean sprouts.
  2. **gomen** I slept at 7:30 **dakara henji shinakatta.** (Srivastava et al., 2020)
     (Translation) Sorry, I slept at 7:30 so I didn't reply.
  3. Good night **dosto ab tumhare hawale ye trend sathiyo** (Srivastava et al., 2020)
     (Translation) Good night, friends! Now the trend depends on you, buddy.

(c) Romanization of Non-English language only:
1. **IMA SUGOKU NOMITAI** (Goh & Wu, 2024)
(Literal translation) I really want to drink that now
2. **Kya biscuit milna bandh hogaya isko** (Srivastava et al., 2020)
(Translation) Has he stopped getting biscuits?
3. **Pahale to aapko modi ji kaam nahi karne de rahe hai**
(Srivastava et al., 2020)
(Translation) First of all, Modi-ji is not letting you work.

(d) Both English and non-English written in non-Latin script:
1. 私1:30ごろに一誠ピックしに行かなきゃいけないみたいで (Goh & Wu, 2024)
(Literal translation) It seems like I have to go to pick (up) Issei around 1:30.

In (a), each language is written in its respective writing system. This approach requires writers to switch keyboards, but the boundary between each language is obvious due to the different scripts. In contrast, in (b) and (c), the entire sentence is written in the Latin script, causing non-English words to be represented in scripts different from their original ones. For example, the Japanese words in sentences (b1) and the Hindi words in sentences (b3) are romanized, while an entire Japanese sentence in (c1) and Hindi sentences in (c2) and (c3) are romanized. Finally, in (d), the entire sentence is written in a non-Latin script, resulting in the English word 'pick' in sentence (d1) being written in Katakana, a component of the Japanese writing system typically used for loanwords.

In (b), (c) and (d), readers must first determine the language of each word and then decode words that are written in unconventional ways, which increases the difficulty and burden for readers. In both Chinese-English and Japanese-English code-switching, these styles occupy low percentages. Goh & Wu (2024) reported that approximately 20% of collected messages were written in these styles, while 80% were written in mixed scripts. This finding aligns with the results of Lin (2005), which indicated that among the 61% of participants who reported writing bilingual messages, only 11% used romanization. Furthermore, if the rules for converting words from one language into the writing system of another are not well established, this task becomes significantly more challenging. The romanization of Japanese and Chinese is relatively straightforward because the romanization conventions are well established. Conversely, the rules for writing English words in Japanese or Chinese are less standardized, which can make it more difficult for readers to understand. Consequently, this makes code-switching of type (d) even rarer; this type was reported only in Goh & Wu (2024) among the three studies mentioned here.

In contrast, this phenomenon—refraining from romanization and preferring script mixing—was not observed in Hindi-English code-switching. Srivastava et al. (2020) reported that among code-switched tweets, texts entirely in the Latin script (styles b or c) were the most frequent (91.5%), followed by the mixed scripts (8.1%) and then Devanagari—the standard script for Hindi— only (0.4%).

Interestingly, all studies reported that inter-sentential code-switching was the most frequent type. This resulted in frequent script-switching within a sentence for Chinese-English and Japanese-English bilingual writers, who most often wrote in a mixed-script style. For example, sentence (a3) from Gonzales & Tsang (2023) exhibited four instances of code-switching, alternating between Chinese and English in the pattern Chinese-English-Chinese-English-Chinese. Switching between two different scripts required changing keyboards, but this technical inconvenience did not appear significant enough to impede script-switching.

Then, why do Chinese-English and Japanese-English bilingual writers prefer mixing scripts despite the technical inconvenience, whereas Hindi-English bilingual writers often rely on romanization? One reason may relate to semantic ambiguity. Chinese of course exclusively uses Chinese characters in its writing system, and Japanese makes heavy use of them. As with Korean, as discussed in Section 7.4, these languages have many homophones, which can increase the ambiguity of words when they are romanized. In contrast, the romanization of Hindi may be less likely to introduce ambiguity insofar as Hindi is natively written in a segmental script, and romanization simply involves the substitution of another segmental script. However, this cannot be the whole story, since as we noted in Chapter 6, writers often use informal romanizations that obscure distinctions found in the native-script spellings, so some ambiguity will inevitably still be introduced. The other factor, then, is familiarity especially when it comes to typing in the Latin alphabet. As noted in Chapter 6, and further discussed in Section 7.3, South Asians are relatively familiar with romanization in their languages, and furthermore often find it easier to type in Latin script than in the native script. This would in turn encourage writers to accept romanized Hindi. In conclusion, it is worth noting that bilingual writers may have different preferences for coding styles, depending on the language pairs involved. These preferences can be influenced by various factors, including social, technical, and readability considerations.

Functional analysis of code switching has revealed that code switching in CMC serves various pragmatic functions. Based on the analysis of self-reports, Gonzales & Tsang (2023) found that participants used code-switching to fill conceptual gaps when expressing concepts available only in one language, to engage in socialization and express intimacy, and to demonstrate their membership in a particular social group. Meanwhile, Nakamura (2020) reported that code-switching was used for expressing emotions, marking humor, emphasis, quoting others' speech, and changing topics. More recently, Montes-Alcalá (2024) analyzed the function of code-switching using a large SMS corpus. The study examined approximately 3,000 SMS messages from young Spanish-English bilinguals in the U.S. and found that code-switching was used for similar functions discussed in Gonzales & Tsang (2023) and Nakamura (2020). None of these functions are unique to CMC; they are often identified as key functions of oral code-switching as well.

Some code-switching in CMC may serve unique functions specific to the medium. Bautista (2004) hypothesized that some code-switching in CMC might occur to deliver messages in the quickest and most convenient way while minimizing effort and time, due to constraints such as character limits in mobile phone texting or time limitations in group chats. Specifically, bilingual or multilingual writers might choose the shortest words and phrases from each language and combine them to create the most efficient (i.e., shortest) messages. Since Bautista (2004)'s proposal, the concept of code-switching for "communicative efficiency" in CMC has been explored by various researchers. Barasa (2016) supported this hypothesis by presenting examples from her dataset where code-switched messages were shorter than either of the monolingual versions.

However, several studies have challenged this hypothesis. For instance, Carrier & Benitez (2010), after analyzing SMS messages from 26 Spanish-English bilinguals, found that messages containing code-switching were actually longer than those written in a single language. Furthermore, in an experimental study where participants were instructed to write concise messages, the results showed no significant difference in message length between the bilingual and monolingual groups.

Barasa (2016) argued that communicative efficiency cannot be measured solely by the character length of a message. For instance, many devices today are equipped with auto-fill functions, allowing writers to complete long words with minimal typing effort, which means that some lengthy messages may require much less typing than their actual length suggests. To further test this hypothesis, employing a more precise method for estimating writing effort and conducting a larger-scale data analysis would be beneficial, rather than relying on the limited sample examples as in Bautista (2004) and Barasa (2016).

In the studies mentioned above, the participants were bilingual writers proficient in both writing systems. However, not all writers share this experience. Heritage language learners, in particular, often encounter unique challenges when trying to master the writing system of their heritage language, especially if it differs significantly from the writing system of the dominant language. For example, Chinese-English heritage speakers, whose dominant language is English, may find the logographic nature of Chinese characters to be a significant obstacle in their script acquisition. This results in significant gaps between speaking/listening skills and writing/reading skills (Wiley, 2017). These speakers have the option to write either predominantly in English or in Chinese using the pinyin system, which is the most common romanization system for Standard Chinese. Zhang (2004) introduced pinyin-based online chat among Chinese-American heritage learners as part

of a college-level Chinese language class and found a positive effect on the acquisition of Chinese characters. However, since the class was at the beginner
level, it was unclear whether the participants could communicate complex
ideas through pinyin-based messages.

7.6 Summary: Factors influencing the selection of writing systems

In this chapter, we examined various writing scenarios involving multilingual
speakers. These speakers use multiple languages as resources in their writing,
even if some languages lack standardized writing systems. We analyzed the
factors influencing the selection of languages and writing systems.

In Section 7.2, we discussed diglossic societies, where speakers communicate in both standard and its local dialects, but only the standard language has a writing system. Through historical and contemporary examples,
we found that writers often adapt the writing system of the standard language for local dialects. Without established orthographic rules, writers may
encounter difficulties, especially when the standard language's orthography
does not suit local dialects. The example of romanized Arabic shows that
spelling conventions can be developed and shared through online communication without official announcements or education. Similar patterns emerge
in India due to high script diversity, as discussed in Section 7.3.

We introduced a digraphic society in Section 7.4: pre-modern Korea. Pre-
modern Korea was a monolingual society, but it employed two writing systems, each used in different social contexts. We showed the process by which
the society transitioned into a single writing system society. During this process, various patterns of code-switching emerged according to the intended
target audience. We also showed that this transition could be expedited by
political factors, as seen in the comparison between North Korea, which experienced a rapid transition, and South Korea, which underwent a relatively
long transition.

Finally, we investigated the code-switching patterns of bilingual writers
in Section 7.5. A comparison among Hindi-English, Chinese-English, and
Japanese-English bilingual writers revealed that these writers prefer different
coding styles depending on the language pairs involved. The choice of script
may be influenced by technical, social, and readability considerations. Once
again, various factors involving the writer, the script, the language and the
language-user conspire to affect what appears on the page.

Chapter 8
Writing in another language

8.1 Prelude

We will begin this chapter by discussing various writing scenarios. The first scenario is writing a project proposal in your native language. (Hereafter, we will abbreviate native language as L1 which is the first language of the person.) As a software engineer, you are writing a proposal for a new game to help children learn to write. In order to complete this task, you will create ideas and organize them into a physical or mental outline. Next, you will convert these ideas into complete sentences, forming a text. Finally, you will review and revise the text to correct errors and enhance its fluency and effectiveness. During this process, to efficiently create a high quality document, you might use various assistive writing tools. For instance, you may use ChatGPT to find new ideas at the initial stage, or use grammatical error correction systems to efficiently proof-read the initial draft.

If you live in a foreign country and need to write this proposal in a language other than L1, you will follow a similar process, but you may struggle in the sentence generating process as a non-native writer. You might frequently use online translation systems such as Google Translate to find suitable words or phrases, or even write sentences in your native language and then translate them. In such a scenario, you may rely more on tools that assist with sentence formulation than when writing in your L1.

If you are a student learning a new language and writing this proposal for a class assignment, your goal is to practice writing in the new language. Your teacher may have a concern about over-reliance on writing assistance tools and prohibit using them. As a result, you may not be able to use the same tools when writing to learn as when writing for other purposes. These scenarios illustrate that writers may use different tools at various stages of their writing process depending on their goals and situations.

This chapter explores how writing assistance tools can help writers at their different writing stages. We will particularly focus on writers who write in the

B. Roark et al., *Tools of the Scribe*,
https://doi.org/10.1007/978-3-032-00831-2_8

second language that they learned after their native language[1]. (Hereafter, we will abbreviate the second language as L2.) To understand how L2 learners write and what are the primary differences from an L1 writing process, we will start with a brief overview of writing process models.

8.2 Writing process models

8.2.1 L1 writing process models

Researchers in the writing education area have collectively shown a strong interest in the writing process. Murray (1972), in particular, criticized final product-focused writing education, and argued for the importance of process-oriented education, which could help students learn how to solve problems during writing, resulting in learning how to write effectively. In this discussion, he proposed pre-writing, writing, and reviewing as the three main stages of writing. Although Murray's three-stage process was accepted early on by writing researchers, a fully developed model that includes a detailed description of each stage and its sub-processes, and interactions between stages, was first proposed by Hayes & Flower (1980).

The overall structure of this model is presented in Figure 8.1. The model consists of three sub-components: the writer's long-term memory, where relevant writing knowledge[2] is stored; the writing task, which encompasses everything external to the writer; and the writing process, which includes planning, formulating[3], and reviewing.

Let us discuss this process in detail using an example. We will go back to the scenario from the last section and start writing a proposal about the new game in L1. The game teaches students the structure of various text genres by having them arrange pre-generated sentences in the correct order for each genre. First, you will start with planning, which involves creating a writing plan and materials given the task. You will retrieve relevant topics from long-term memory and then organize these ideas in an appropriate order. Before arranging them, you may evaluate each idea to determine its

---

[1] The major points of this chapter—such as the use of writing assistant tools and their impact on writers—are applicable to writers who compose in foreign languages beyond their second language. These writers share core processes with L1 and L2 writers, but they face more challenges during their writing, such as greater cognitive demands.

[2] Examples of knowledge include writing-related aspects, such as the structure of specific text genres or grammar, as well as topical knowledge, such as history, society, and technology.

[3] The original terminology used by Hayes & Flower (1980) is 'translating'. We replace 'translating' with 'formulating' to avoid confusion when 'translating' refers to 'translating one language into another' later in this chapter. The term 'formulating' was used by Zimmermann (2000) for the same reason.

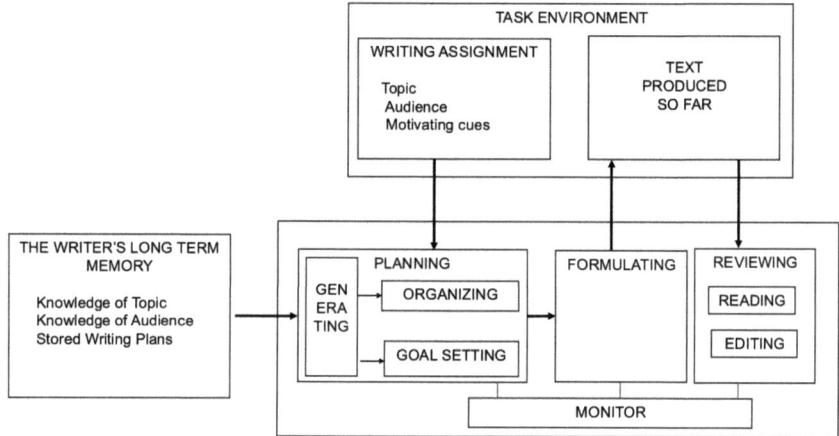

Fig. 8.1 Writing process model, redrawn from Hayes & Flower (1980).

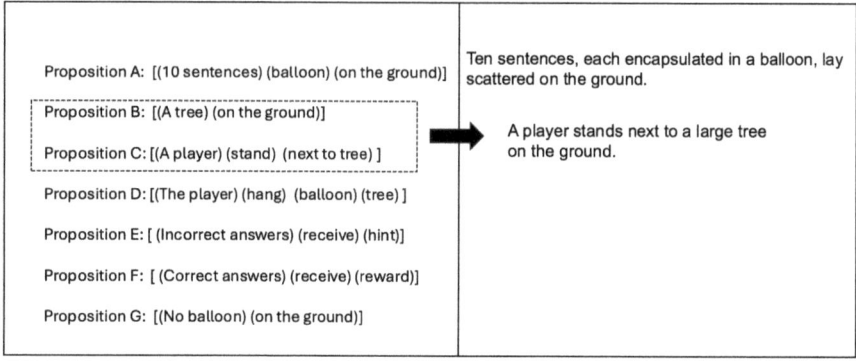

Fig. 8.2 Formulating the writing plan (left box) into sentences (right box). Source: own work

usefulness, removing any that are not useful. You may place ideas in a hierarchical order or in a temporal order. At this stage, the materials are in the form of propositions or concepts, not yet in language form.

Next, you will move to the formulating process in Figure 8.1. Figure 8.2 illustrates an example of the formulating step. You will take each proposition from the left box and convert it into a complete and well-formed sentence in your L1, step-by-step. For instance, you take proposition A and convert it into "Ten sentences, each encapsulated in a balloon, lay scattered on the ground." You then take the next proposition and generate a sentence. You repeat this process until all propositions are formulated into complete sentences.

After the formulating process, the final process is reviewing, where you examine the text, identify areas of sub-optimal quality, and make improvements by correcting errors or enhancing connections.

How do these processes interact with each other during writing? Hayes & Flower (1980) introduced a 'monitor' to manage the interactions among these processes. Generally, the writing process follows the order of planning, formulating, and reviewing, but the detailed processes may vary according to individual writing styles. These individual differences are reflected during goal setting in the planning step and influence the interactions among processes in turn. For example, a breadth-first writer may set a goal to draft a working version first and perfect the details later. They follow a sequence of generating, organizing, formulating, and reviewing with few interruptions, addressing most issues in the previous processes only after completing the entire process. Conversely, a depth-first writer focuses on perfecting each sentence before moving to the next, consistently alternating between formulating and editing. The configuration stores the individual's goals and writing style, and the monitor controls the interactions among processes based on this configuration.

In addition, Hayes & Flower (1980) argued that among all sub-processes, the generating and editing sub-processes occur relatively freely and are likely to interrupt other processes. For instance, while writing about proposition E in Figure 8.2, you may realize there is no idea about 'hint'. You may then stop the formulating process and go back to the generating process to retrieve ideas about the hint. In another example, while writing the next proposition, you may notice spelling errors in the previous sentence you already wrote, causing you to move to the reviewing process to correct the spelling errors.

8.2.2 L2 writing process models

Now, let us examine the writing process of L2 writers. Unlike its L1 counterpart, L2 writing involves an additional language (L1) other than the target language (L2) in which the final text should be written. The switch between L1 and L2 happens frequently during the writing process (Cumming, 1989; Manchon et al., 2000; Woodall, 2002), and the L2 writing process model needs to incorporate these differences. Börner (1989) and Zimmermann (2000) developed L2 models by augmenting Hayes & Flower (1980)'s L1 model and integrating L2 specific characteristics. In this chapter, we will focus on Zimmermann (2000)'s model because it provides detailed insights into the formulating process, where the major differences between L1 and L2 lie. Additionally, the model is based on an analysis of the actual writing processes of L2 writers. A diagram of Zimmermann (2000)'s L2 writing process is provided in Figure 8.3.

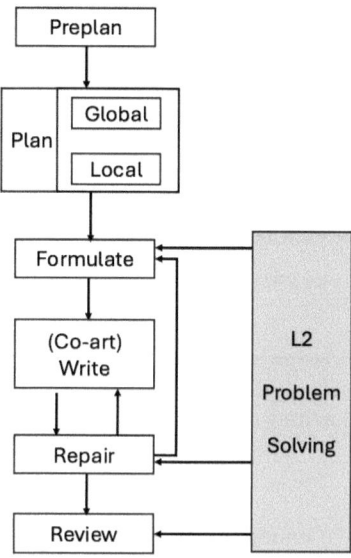

Fig. 8.3 L2 writing process model, redrawn from Zimmermann (2000).

We will revisit our example of writing a proposal for a new children's game. This time, however, the proposal will be written in L2. As a Korean engineer, you will write this proposal in English. You will begin by generating and organizing your ideas. Similar to writing in your L1, you may choose to use only English, the target language you are writing in. In other cases, you might use only Korean, or switch between English and Korean. Mixing the two languages could result in an outline that includes both languages, as shown in Figure 8.4. In this example, the proposition [(10 sentences) (balloon) (on the ground)] is replaced by [(10 sentences) (풍선) (on the ground)], using the Korean word 풍선 ⟨pungseon⟩ for 'balloon'.

The choice of language during planning is influenced by various factors such as the task complexity, L2 proficiency, and personal preference (Qi, 1998; Manchon et al., 2000). Despite concerns of language teachers that using L1 during the writing process may hinder L2 writing development, Manchon et al. (2000) found that language switching occurs naturally and automatically, without requiring conscious effort. As a result, it does not diminish the quality of the L2 outputs.

Next, you will formulate each proposition in your plan into English sentences. During this step, you may encounter more challenges than when writing in your L1 due to limited knowledge of L2. To address these issues, an L2 problem-solving module—one of the key differences between L1 and L2 models—operates alongside the overall writing process. This module is activated during the formulating and reviewing processes, performing various

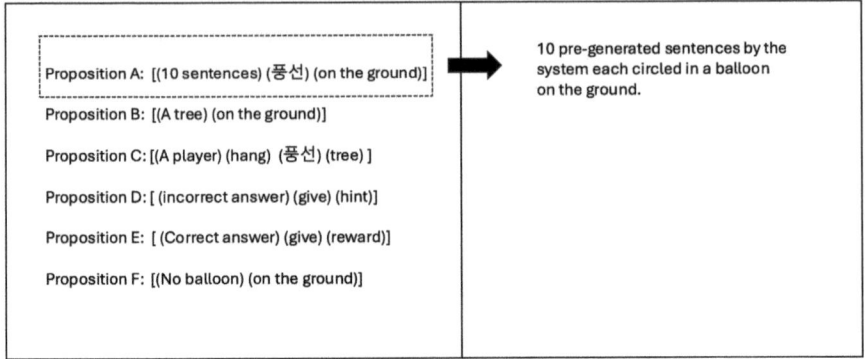

Fig. 8.4 Formulating the writing plan (left box) into L2 sentences (right box). Source: own work

linguistic tasks to resolve issues encountered during L2 sentence construction, such as vocabulary retrieval and grammar rule application.

We will further explore the formulating process using Zimmermann (2000)'s model presented in Figure 8.5. This model provides many more details about the L2 formulating process.

Let's work on the proposition "[(10 system sentences) (풍선) (on the ground)]" and formulate it in English. First, "(10 system sentences)" is formulated into "10 pre-generated sentences generated by the system", which corresponds to TentativeForm$_1$ in Figure 8.5. This TentativeForm$_1$ can be slightly modified (TentativeForm$_1$/modified) and evaluated.

During the evaluation process, you might find this form too long and repetitive due to the repeated word "generated." Therefore, you decide to postpone the decision and create a TentativeForm$_2$, that is "10 pre-generated sentences by the system" After the evaluation step, you decide to accept TentativeForm$_2$ and finally go through a 'co-articulation' step in Figure 8.3, wherein a mental formulation of a sentence is converted into graphical symbols on a physical medium, such as paper or a computer screen, usually by typing.

The next part of the proposition is 풍선, which is in L1 and needs an appropriate L2 equivalent. This step is highlighted by the grey boxes in Figure 8.5. The TentativeForm$_{L1}$ '풍선' goes through the L2 Problem Solving module and is formulated into 'a balloon'. The use of L1 is a unique aspect of L2 writers, where TentativeForm$_{L1}$ serves as an intermediary stage. However, Zimmermann (2000) found that this overt translation step does not frequently appear in the L2 writing process, and the occurrences further decrease as a writer's L2 proficiency increases.

After L2 problem solving, the TentativeForm$_1$ form "each circled in a balloon" will be formulated and then undergo an evaluation process. During the evaluation, you might question the accuracy of the phrase. The original in-

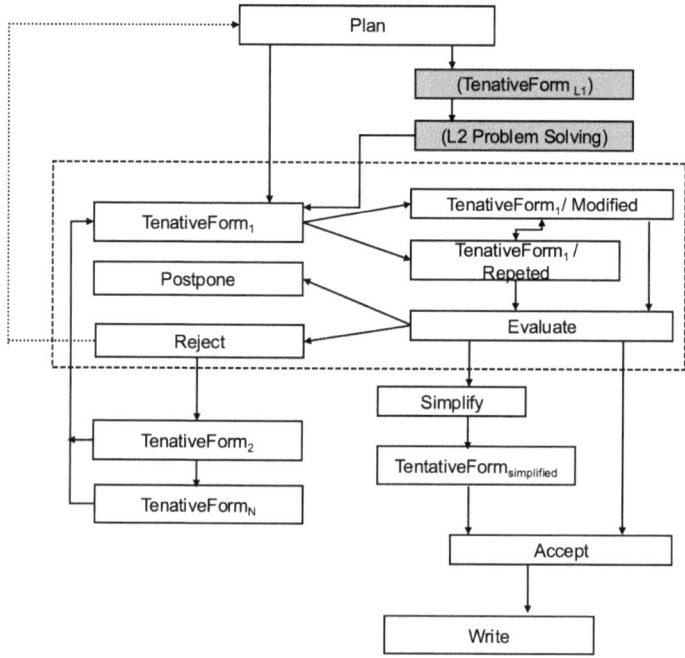

Fig. 8.5 Formulating process in L2 writing process, redrawn from Zimmermann (2000).

tention might have been "each encircled in a balloon" or "each encapsulated in a balloon," but you may struggle to find the exact words in L2. Eventually, TentativeForm$_1$ will undergo a 'Simplify' step, which simplifies the idea, creating a simpler language form, resulting in "each in a balloon". This simplified version is accepted and also proceeds to a 'co-articulation' step.

Zimmermann (2000) found that the formulating process is highly iterative, with certain sub-process sequences occurring frequently. The most common sequence is TentativeForm → evaluate → accept → co-articulate → review. This sequence closely resembles the typical process of L1 writers. However, he also found that the L2 writing process has several characteristic differences from L1. First, there are unique steps—TentativeForm$_{L1}$ and the L2 Problem Solving module—in L2. Secondly, the 'Simplify' step occurs in both L1 and L2 writing, but is more common in L2 writing. This is because limited knowledge and resources in L2 often lead to a compromise in the goal and a simplification of the output message. Finally, the L2 writing process is marked by more frequent revisions. Zimmermann (2000) observed that the participants in their study revised about 1.3 times more when writing in their L2 compared to their L1. Additionally, these revisions often involved technical corrections, such as fixing grammar or spelling errors. These differences can influence the writing aids that L2 writers select, often leading them to prefer tools that differ from those chosen by L1 writers. We will delve into this topic in detail in the upcoming sections.

8.3 Writing assistance tools and the writing process

Next, we will discuss how both L1 and L2 writers use writing assistance tools in their writing process. However, before diving into the writing process itself, we will first explore the classes of existing assistance systems. Advances in natural language processing (NLP) and artificial intelligence (AI) have led to the development of various assistance systems for writers. These systems— some of which we touched on briefly in prior chapters—range from traditional spelling and grammar correction systems to emerging large language models (LLMs). They have significantly changed the writers' writing process and greatly improved their writing efficiency. The use of these systems has become essential for many writers.

8.3.1 Writing assistance systems

Commonly used writing assistance systems can be categorized into four groups as in Alharbi (2023): automated writing evaluation (AWE) systems, automated writing corrective feedback (AWCF) systems, machine translation (MT) systems, and large language models (LLMs) such as ChatGPT. AWE and AWCF systems were originally developed to assist the writing process, while MT systems and LLMs were developed for other purposes or more general use, but they are often used to help with writing as well.

AWE systems, such as Educational Testing Service's Criterion Online Writing Evaluation Service, Vantage Learning's MY Access!, or Pearson's WriteToLearn, provide various aids on both higher order concerns like content and organization, as well as lower order concerns such as spelling, grammar, and conventions. These systems are primarily designed for writing with a learning purpose, ensuring appropriate assistance for common academic genres like argumentative and narrative essays. One standout characteristic of AWE systems is the diagnostic feedback they offer when the initial draft is ready. This feedback includes holistic proficiency scores, generated via automated scoring engines that are trained on a vast collection of essays rated by humans and machine learning algorithms. In addition to the holistic scores, it can provide analytical scores that give insight on sub-areas such as language use, coherence, and content. Writers can use this feedback to identify and enhance problematic areas in their writing. For example, writers receiving a low coherence scores can examine and improve elements related to coherence, such as transitional words and pronouns. However, these scores are not generic; they are valid only for the specific genre of essays that the scoring engines have been trained on. Consequently, they cannot be applied to other styles of writing tasks.

Unlike AWE systems, **AWCF** systems primarily focus on lower order concerns. They are the most familiar and ubiquitous among the four groups. A well-known example is the spelling and grammar correction function integrated into word processors such as Microsoft Office. Also, Grammarly and ProWritingAid are popular systems in this group.

MT systems are unique tools for L2 writers, as they are not typically used by L1 writers. A growing number of L2 writers now choose to compose content in their native languages, leveraging translation systems, and subsequently refining the output through post-correction. Based on multiple survey studies conducted between 2010 and 2020, L2 writers' MT usage has been consistently increasing, making these systems common tools in writing tasks today. O'Neill (2019a) reported survey results from more than 300 L2 students in a US university, and found that approximately 90% of L2 students at the university level use MT for their writing.

LLMs have appeared only recently, but they already have had a strong impact that has transformed the educational field. ChatGPT, OpenAI's virtual assistant based on LLMs, is one of the most well-known and accessible LLMs. Since its launch in November 2022, ChatGPT has had a tremendous impact on students' writing processes. According to the online survey conducted in February 2024 of 588 U.S. college students (Intelligent, 2024), 96% of participants used it for at least one school project during the current academic year, with 37% being consistent users. The most common usage was for writing assignments (69%) which highlighted the widespread use of LLMs in writing tasks. Unlike other systems designed to assist a few specific tasks, LLMs can provide assistance throughout entire writing processes, and NLP researchers and language educators have explored deploying LLMs as writing tutors for a range of tasks. These tasks include assigning holistic proficiency scores as diagnostic feedback (Yancey et al., 2023; Yoon et al., 2023), offering corrective feedback on grammatical errors (Wu et al., 2023; Coyne & Sakaguchi, 2023; Fang et al., 2023), providing explanations about the characteristics and requirements of the target genre, and generating ideas related to the topics.

In summary, writers can employ these systems for a wide range of subtasks related to the writing process. Table 8.1 outlines how these tools can be utilized at each stage of the writing process.

It is important to note that these usages are based on survey studies about the students' writing process (Clifford et al., 2013; O'Neill, 2019a; Burstein et al., 2018) and classroom writing studies involving teachers' interventions (Su et al., 2023; Yan, 2023; Liu et al., 2024; Levine et al., 2024). In particular, studies on the use of LLMs are still underexplored, and some of the usages are based on recommendations from researchers and teachers rather than on how actual writers use them (Atlas, 2023; dos Santos et al., 2023). Consequently,

Tool type	Process	Usage
AWE	Reviewing	Identify areas for further improvement based on numeric scores or linguistic features (e.g., analytic scores), provide corrective feedback on spelling, grammatical errors, and conventions
AWCF	Reviewing	Provide corrective feedback on spelling, grammatical errors, and conventions, suggest alternative words, adjust style, tone, and politeness
MT	Formulating	Provide similar function to an online dictionary: translating individual words, phrases, sentences (from L1 to L2)
	Reviewing	Back-translate what students wrote for double checking (from L2 to L1)
	Planning	Provide explanations about the writing task (e.g., genre structure), provide relevant information about the topic, generate ideas, summarize literature
LLMs	Formulating	Formulate sentences given the ideas in the words or phrases, compose paragraphs or an entire essay given a topic or an outline, create example sentences as references (e.g., examples of persuasive opening sentences), and all usages mentioned in MT.
	Reviewing	All usages mentioned in AWE, AWCF, MT

Table 8.1 Usage of writing assistance tools during writing process

Table 8.1 is not comprehensive; it may not fully reflect some frequent usages by writers, while some listed usages may not frequently occur during actual writing.

As indicated in Table 8.1, many tools offer support during the formulating and reviewing stages, while assistance in the planning stage is limited. However, the advent of LLMs could offer significant support throughout the entire writing process including the planning stage.

8.3.2 How writing assistance tools are being used in L1

We will now explore how writers utilize these tools in their writing process, focusing on their impact on efficiency, output quality, and learning gains. The recent study by Levine et al. (2024) provides detailed insights into how writers, particularly L1 student writers, use LLMs. Unlike previous studies such as Yan (2023) and Liu et al. (2024), this research did not restrict how students should use ChatGPT, except for prohibiting the creation of an entire essay. This approach allowed for observations that more closely reflect students' actual usage. The researchers recruited 12 high school students in the USA and conducted two 60-minute after-school writing sessions. During these sessions, pairs of students collaborated to write an argumentative essay. The study analyzed the types of questions students asked ChatGPT and how they incorporated ChatGPT's responses into their writing. The results indicated that students used ChatGPT most frequently during the planning stage (81%), followed by reviewing (31%) and formulating (16%), suggesting that LLMs can provide effective support during the planning stage.

At the planning stage, students primarily used ChatGPT to gather relevant information and evidence to support their arguments. During the formulating stage, some students used a novel approach by asking ChatGPT to provide examples of sentences with specific rhetorical functions, such as a very persuasive opening statement. In both cases, students refrained from directly copying ChatGPT's output; instead, they developed their own ideas and sentences based on ChatGPT's responses. Finally, during the reviewing stage, students asked ChatGPT to correct grammatical errors in their writing or to revise it for smoothness and consistency. Unlike their creative and proactive approach during the planning and formulating stages, students accepted or rejected the edits as a whole, rather than reviewing them individually to learn from their mistakes.

The Levine et al. (2024) study is largely consistent with the results of the online survey mentioned above (Intelligent, 2024). Among students who used LLM for writing, the majority (75%) used it for idea generation. In addition, they frequently used it for reformulating sentences and proof-reading. However, a significant number of students generated more than just sections using LLMs. Among the students who used LLMs for their writing assignments, 29% reported using them to write entire essays, even though this goes against the instructions from their professors or institutions. These results highlight that LLMs can be a double-edged sword as a writing assistance tool. LLMs can be valuable writing assistants, offering tailored help and exposing students to diverse examples that teachers may not easily provide due to limited resources. While LLMs can increase writing efficiency, they do not always lead to learning gains; careless use may even hinder a writer's learning progress. Therefore, it is crucial for writers to use these tools with caution, keeping their writing goals in mind. When writing for learning purposes, they should avoid using LLMs as a cheating tool or relying on it to do their work for them. However, when writing for other purposes, this caution may not always be necessary. In such cases, it is reasonable for writers to develop strategies to maximize the writing efficiency and output quality by incorporating LLMs into their writing process, even using them as co-writers.

8.3.3 How writing assistance tools are being used in L2

We now turn our focus to how L2 writers get help from these tools during their writing. Since the main difference between L1 and L2 writing processes lies in formulating stages, with L2 writers experiencing more frequent and serious problems in lower order concerns, we will focus on assistance with those

Type	Feedback Sentences
Direct feedback	Online class(use 'classes') are not like regular class you take.
Indirect feedback	Online class(?) are not like regular class you take.

Table 8.2 Examples of direct and indirect feedback for lower order concerns

Type	Input	Output
Translating from L1 to L2	"Online classes are not like regular class you take"	Las clases en línea no son como las clases regulares que tomas
Formulating sentences given words	10 sentences in a bubble on the ground	Ten sentences, each encapsulated in a bubble, appear on the ground.
Formulating example sentences	A hook for argumentative essays regarding change in school mascots	Imagine cheering for a mascot that doesn't reflect or respect your school's identity—how would that feel?
Rewriting	I agree that providing online or video-based learning at home is useful for students.	There's no doubt that offering online or video-based learning at home is an invaluable resource for students.

Table 8.3 Examples of different types of assistance for lower order concerns

issues from this point forward. While there are variations in the assistance functions offered by different system types, all systems offer support for lower order concerns, as presented in Table 8.1.

Assistance related to lower order concerns is generally divided into direct and indirect feedback, as outlined in Ellis (2009). Direct feedback is a type of feedback where systems provide the correct forms for errors. Conversely, indirect feedback points out the existence of errors, either by indicating their presence or by precisely identifying their location and type, without providing the correction. Table 8.2 provides examples of the direct and indirect feedback. The original sentences are extracted from an L2 writer's essay, and the system assistance is provided within the parentheses. '?' was used as a symbol to indicate the presence of an issue.

Beyond these two types, which were typically used for teacher feedback, MT and LLMs introduced new types of assistance. Table 8.3 shows examples of these new types of assistance generated by ChatGPT.[4]

---

[4] In order to generate the output (system assistance), we used the ChatGPT 4.0 model. We used the following prompts for each type of assistance:

- Translating: Translate the following sentence in Spanish.
- Formulating sentences given a words: Write a sentence using the words in the input; the sentence should be grammatical and complete, within 15 words.
- Formulating exemplar sentences: Can you give me an example of an engaging opening for an argumentative essay about the input?
- Rewriting: Rewrite the following sentence in a persuasive tone.

We submitted the prompt and input to OpenAI Playground: https://platform.openai.com/playground.

The type of assistance significantly influences the role of L2 writers during the revision process. L2 writers receiving indirect feedback must independently identify the actual errors and determine the correct forms, whereas those receiving direct feedback can automatically accept the corrected forms without much deliberation. For instance, writers only need to replace 'class' with the system suggestion, 'classes' when receiving the direct feedback in Table 8.2. However, when receiving the indirect feedback, writers must first recognize there is a noun number error and then find the correct form 'classes' on their own. Therefore, indirect feedback requires more active involvement from L2 writers than direct feedback. Conversely, L2 writers using MT systems may require the least active involvement. They can simply accept the generated sentences by the systems without formulating them by themselves. While the type of feedback does not completely determine the role of L2 writers, it significantly influences the writing process and, consequently, the learning outcomes. We will further discuss this aspect towards the end of this chapter.

How, then, do L2 writers actually use these assistance systems during their writing process? Generally, L2 writers' reactions can be divided into two groups: a) responding to the feedback and b) rejecting the feedback. Responding to feedback includes either accepting it or making their own revisions, such as correcting errors in their own way or removing entire sentences that contain errors. Rejecting feedback means that writers make no revisions or simply ignore the feedback. Various factors influence this decision, which can be related to individual writers and other contextual elements (Karatay & Karatay, 2024).

8.3.3.1 Use of AWCF systems

First, we will focus on AWCF systems, particularly Grammarly. Grammarly is a free AI writing assistant known for its high performance and widespread use among both L1 and L2 students. Additionally, its usage has been relatively well studied in recent L2 writing research. Grammarly provides both premium and free versions.[5] It offers a user-friendly interface for text submission, allowing users to either paste their text or type directly into an interactive window. For the submitted text, it jointly provides both indirect and direct feedback. Initially, it highlights errors with underlines without offering corrections, enabling users to attempt to correct their mistakes independently. (Hereafter, we will refer to a highlighted area by the system as a 'flagging'.) Users can also view suggested corrections by hovering over these underlines. Finally, it offers an option called "correct with assistant," which provides detailed explanations of the errors along with the corrections.

---

[5] It is accessible at `https://www.grammarly.com`.

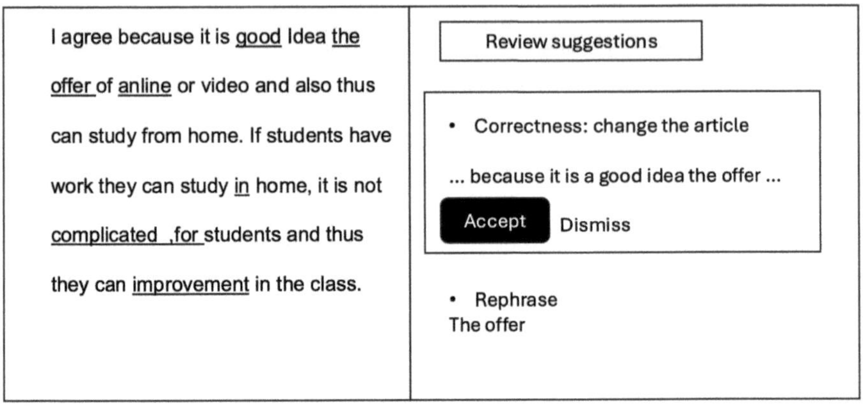

Fig. 8.6 An example of Grammarly feedback. text window (left box) and the system feedback (right box). We used a stylized version instead of a picture of the actual Grammarly interface. Source: own work.

Figure 8.6 illustrates an example of Grammarly's feedback for the introduction of an essay about online classes.[6] The left box in the figure is the text window where students can type or copy and paste their input text. The right box displays the corrections the system offers.

The words with grammatical or other errors are marked with underlines in the left box, while the explanation of each error is presented in an itemized list in the right box. When one clicks on an item, the system provides an explanation on how to fix it, along with accept or reject buttons. The right box can be hidden from the display to allow users to identify errors in the left window and try to make their own corrections instead of directly accepting the system's suggestions.

Beyond just error corrections, the premium version of Grammarly offers feedback on a variety of issues including style, word choices, sentence variety, politeness, and inclusive language.

However, feedback from AWCF systems does not always lead to writers' revisions. One crucial factor determining the writers' response is their trust in the AWCF system, which is influenced by its accuracy. Accuracy can be formally evaluated using two key metrics: precision and recall. Precision, involving false-positive errors, measures the proportion of detected errors that are actually true errors. Recall, relevant to false-negative errors, measures the proportion of total errors present in the text that were successfully corrected by the system. If the precision is low, then a high proportion of the errors detected by the system are not actual errors (over-correction) or the suggested corrections are incorrect. If recall is low, the system fails to detect many errors

---

[6] Due to permission issues, the authors created a sample essay that mimics common errors made by L2 writers, rather than using an authentic essay written by an L2 writer.

in the writer's text. High precision and recall ensure that the corrections suggested are both accurate and comprehensive.

Ranalli (2022) investigated the accuracy of Grammarly's feedback on essays written by tertiary-level L2 English learners. First, he focused on evaluating two aspects of precision: a) whether a flagging involved a true error, and b) whether the proposed correction was accurate. He submitted 68 essays and received approximately 1500 flaggings. Each flagging was classified into four types: 'accurate', 'inaccurate', 'indeterminate', and 'generic'. 'Accurate' were cases where the flaggings included true errors and Grammarly provided accurate corrections that could fix specific errors. 'Inaccurate' were cases where either the flaggings did not include true errors or the Grammarly's corrections were inaccurate. 'Indeterminate' were cases where it was difficult to decide whether the flaggings were correct or not. These involved feedback about style, word choices, sentence variety, politeness, and inclusive language. Finally, 'generic' were cases where the system provided only generic approaches without specific corrections. He evaluated the precision after removing 'indeterminate' and 'generic' types from the dataset. Grammarly demonstrated high performance in both aspects of the precision measure, achieving 0.88 and 0.81, respectively.

Next, he evaluated recall for four common L2 error types: spelling, articles, prepositions, and subject-verb agreement. On the same dataset mentioned above, he annotated all errors that fell into these four categories, regardless of whether they were detected by Grammarly or not. Overall, recall was quite low. The highest recall was for spelling errors at 0.86, followed by subject-verb agreement at 0.67. Recall for both article and preposition errors was below 0.50, indicating that the majority of these errors went undetected. In summary, Grammarly was highly accurate for the errors it detected, although many errors in L2 writers' texts remained undetected.

Despite the low recall, Grammarly's performance was significantly better than many available AWCF systems. For instance, traditional AWCF systems based on grammar rules and automated syntactic analysis generally achieve even lower recall. The same study found that Microsoft's legacy grammar error correction system, MS-NLP, could detect ten times fewer errors than Grammarly, although the precision of both systems was comparable.

Three recent studies (Koltovskaia, 2020; Thi et al., 2023; Ranalli, 2021) investigated detailed analyses of L2 writers' reactions to Grammarly's feedback. Koltovskaia (2020)'s study was conducted as part of an ESL writing course at a university in the U.S. Two participants, one advanced and one intermediate learner, were selected for the study. They submitted their drafts to Grammarly, and the feedback was directly displayed and editable during the revision. Similarly, Thi et al. (2023)'s study was part of an EFL writing course at a university in Hungary. Thirty-one Hungarian students at the intermediate proficiency level participated in the study. Finally, Ranalli (2021)'s study was conducted outside of a classroom setting. This study recruited three advanced and three intermediate students from a university in

Paper	N. of Participants	Input length	N. of Flaggings	Flaggings per 100 words	Rejection rate (%)
Koltovskaia (2020)	2	862	40	2.3	23
Ranalli (2021)	6	751	313	6.9	34
Thi et al. (2023)	31	Unspecified	481	Unspecified	71

Table 8.4 Grammarly's flagging frequencies and rejection rate

Paper	Proportion of each type in the total flaggings (%)			Rejection rate (%)		
	accurate	inaccurate	indeterminate	accurate	inaccurate	indeterminate
Koltovskaia (2020)	80.0	20.0	0.0	25.0	13.0	0.0
Ranalli (2021)	40.0	8.0	52.0	5.0	56.0	53.0
Thi et al. (2023)	44.1	5.6	50.3	72.0	81.0	70.5

Table 8.5 Distribution of the feedback types and rejection rates

the U.S. All participants, except one, had used Grammarly. The participants were instructed to address all flagged issues, and in cases where they chose to ignore a suggestion, they were asked to formally select the "Ignore" button. Both Koltovskaia (2020) and Ranalli (2021) requested students to record a video of their computer screens to capture their revision process, while Thi et al. (2023) did not.

Table 8.4 provides information about the input texts and frequencies of Grammarly's flaggings. It also includes the average rejection rate, which is calculated by dividing the total number of rejected flaggings by the total number of flaggings. The rejection rate is an important metric that shows to what extent users are engaged with the feedback.

The rejection rates for Koltovskaia (2020) and Ranalli (2021) were similar, with both being under 50%. However, the rejection rate for Thi et al. (2023) was notably high, more than three times higher than that of Koltovskaia (2020). This result indicates that L2 writers in Thi et al. (2023) were generally not engaged with Grammarly and rejected most of the feedback they received.

Thi et al. (2023) argued that the main reasons for writers' feedback rejections were the instructional contexts and participants' distrust of AWCF systems, while Ranalli (2021) attributed them to feedback types. We further investigated whether these factors could explain the higher rejection rates in Thi's study compared to two other studies. To achieve this, we first analyzed the distribution of each feedback type in the total flaggings and the rejection rate for each type, similar to Ranalli (2021). The results are presented in Table 8.5.

The distribution of feedback types was consistent in both Ranalli (2021) and Thi et al. (2023)'s studies, with the indeterminate type being the majority at over 50%. This was followed by the accurate type, while the inaccurate type constituted less than 10%. Koltovskaia (2020) did not receive flaggings in indeterminate type because they used the free version, which did not provide

Paper	Participants			Flaggings	
	ID	Proficiency	Attitude	N	Rejection rate (%)
Koltovskaia (2020)	Kesley	Intermediate	trust	26	16
	Alex	Advanced	distrust	14	43
Ranalli (2021)	Yong	Intermediate	trust	108	32
	Na	Advanced	distrust	43	14

Table 8.6 Background information and total flagging counts of individual participants

ID	Proportion of each type in the total flaggings (%)			Rejection rate (%)		
	Accurate	Inaccurate	Indeterminate	Accurate	Inaccurate	Indeterminate
Kesley	69	31	0	11	13	-
Alex	100	0	0	43	-	-
Yong	35	6	59	0	17	53
Na	44	2	53	0	100	22

Table 8.7 Distribution of the feedback types and rejection rates for the individual users

stylistic or generic suggestions. After ignoring the indeterminate type, the distribution across all three studies was similar. The relationship between feedback types and rejection rates was significant in Ranalli (2021)'s study, where writers mostly accepted accurate flaggings but rejected most inaccurate and indeterminate ones. In contrast, Thi et al. (2023) did not observe this pattern, reporting high rejection rates across all types, each exceeding 70%.

Next, the instructional contexts could not explain the differences between these studies, as both Thi et al. (2023) and Ranalli (2021) emphasized issues related to higher order concerns, such as the content and organization, more than lower order concerns. Therefore, there may be other important factors that contributed to the high rejection rate in Thi et al. (2023).

In addition to these external factors, individual writers' characteristics strongly impact their reactions to AWCF feedback. To further investigate this aspect, we extracted four participants whose follow-up interviews were available from Koltovskaia (2020) and Ranalli (2021) and further analyzed their response patterns. Table 8.6 presents the characteristics of individual participants, including their proficiency level, attitude toward Grammarly, and overall usage patterns of Grammarly.

L2 learners' proficiency levels and their attitude toward AWCF systems have been considered as two critical individual factors that influence the writers' engagement with the feedback. However, it is difficult to find clear patterns from Table 8.6. For instance, the rejection rate for the advanced learner (Alex) was higher than that of the intermediate learner (Kesley) in Koltovskaia (2020)'s study, while the reverse trend was found from Ranalli (2021). Furthermore, in Ranalli (2021)'s study, the learner who expressed distrust to Grammarly (Na) rejected less feedback than the learner who showed trust (Yong). For further examination, Table 8.7 provides the rejection rate of each feedback type for each participant.

Kesley and Yong were both intermediate learners, notable for their low rejection rate of inaccurate flaggings. This was especially evident when compared to Na, an advanced learner who had a 100% rejection rate of inaccurate flaggings. According to Koltovskaia (2020)'s interview, Kesley heavily relied on Grammarly, accepting all flaggings if she did not know related grammar rules. In contrast, Yong was a more confident L2 writer and also an experienced Grammarly user. As noted in Ranalli (2021), Yong understood that Grammarly could make mistakes and didn't blindly accept all flaggings. However, he still struggled to identify inaccurate flaggings and accepted most of them. As a result, Kesley made 23 revisions, with 30% introducing errors due to accepting inaccurate flaggings or making erroneous revisions. Yong made 73 revisions, with only 7% resulting in errors. From the start, Yong received very few inaccurate flaggings, so accepting the majority did not lead to a significant number of new errors.

Alex and Na were both advanced learners, but they differed significantly in their rejection rates of accurate flaggings. Alex rejected 43% of these flaggings, while Na rejected none. According to Koltovskaia (2020), Alex often distrusted Grammarly and only accepted flaggings when he knew the related grammar rules, leading to a high rejection rate of accurate flaggings. The most effective user was Na, who was the most advanced ESL writer, and she could fix all errors identified while not introducing any errors. This was possible because the errors identified by Grammarly were all within her knowledge. During the interview, she mentioned that she did not gain any new knowledge from using Grammarly. Instead, it functioned as a labor-saver by reducing the time and effort she needed to spend on proofreading her mistakes. Accepting Grammarly's feedback automatically, without careful consideration and active verification, was common for most of the participants. This was evident from the minimal amount of time they spent addressing each flagged issue.

Individual usage patterns in these studies suggest that many L2 writers made sub-optimal use of Grammarly's feedback. Low or intermediate learners struggled to distinguish between inaccurate and accurate feedback, leading to the introduction of new errors. Due to improvements in AWCF systems, the proportion of new errors introduced is likely to be low. However, for certain writers, such as Kesley, this can still represent a significant proportion and remains beyond their ability to fix. Conversely, high rejection rates based solely on their own knowledge, without active research, may result in correcting only a subset of their errors, leaving the final output not significantly different from their original texts.

In contrast to three studies discussed above, Barrot (2023) found a positive engagement between L2 writers and Grammarly in his study conducted at an Indonesian university. The study involved 65 ESL students enrolled in two writing classes, with half of them selected as a treatment group to use Grammarly. According to a survey conducted at the end of classes, the participants reported that Grammarly's feedback could help them to identify errors in their texts and deepened their understanding of the grammar. He did not provide a detailed analysis of how participants responded to each flagging they received, making it difficult to determine whether participants could truly distinguish accurate feedback from inaccurate ones as reported in the survey. However, the study offered both quantitative and qualitative analysis of the participants' writings. By comparing the accuracy of texts written before and after training with Grammarly, he found that the training significantly reduced errors related to lower order concerns. Notably, sentences with serious errors that impeded understanding were meaningfully reduced. Although some participants showed an over-reliance on the system and blindly accepted the suggestions, many participants were more active and more engaged, leading to the improvement. Finally, it is worth mentioning that errors were reduced even in writing not using Grammarly, indicating that the impact was not merely from efficiently incorporating Grammarly's feedback but rather from genuine learning gains.

A key difference between participants in Barrot (2023)'s study and the participants from Koltovskaia (2020) and Ranalli (2021) was the level of active engagement. Many participants in Barrot (2023)'s study reported that they read the meta-linguistic explanations and verified the accuracy of the feedback by searching the internet or discussing with friends when they were confused by the explanations. This active engagement not only helped them to improve the quality of their writing but also fostered self-directed learning. Dizon & Gayed (2021) reported similar results in a study targeting EFL writers' mobile texts. They found that writers who received Grammarly feedback achieved both higher grammatical accuracy and lexical richness. These studies support the notion that AWCF systems can improve the quality of L2 writers' output texts, particularly by reducing errors related to lower order concerns.

In summary, L2 writers' engagement with AWCF systems is influenced by a combination of contextual and individual factors. However, it is known that without active engagement, merely passively accepting or rejecting feedback results in little to no learning, and the output texts may show relatively moderate improvement depending on the writers' proficiency level.

Based on these results and their own experiences, many educators suggested further enhancements to make these tools more effective as learning aids. One significant concern is cognitive overload due to the provision of too much feedback at once. AWCF systems return feedback for all detected errors and other stylistic issues simultaneously, resulting in a large volume of feedback. Many previous studies found that this overwhelmed and sometimes even discouraged L2 learners (Dikli, 2010; Barrot, 2023; Koltovskaia, 2023). To address the cognitive overload issue, Roscoe et al. (2014, 2017) proposed a hierarchical presentation of feedback as a design principle for learning aids. In this approach, feedback is prioritized based on its impact and importance, with the system initially presenting the most critical feedback. Upon the learner's request, additional, less critical feedback is then delivered. Simi-

larly, AWCF systems that create an error profile for specific users based on
the analysis of their errors, prioritize error types based on their frequency
and potential for the communication disruption, and provide feedback from
the most important to the least important, will be useful as learning aids. On
the contrary, when AWCF systems are used as proofreaders, it is important
to provide comprehensive feedback for all existing errors to support efficient
and speedy proofreading. Therefore, it may be worthwhile to consider a dual
design for different purposes, delivering feedback in ways that depend on the
writer's needs.

8.3.3.2 Use of LLMs for writing assistance

We will next discuss grammatical error correction using LLMs, specifically
ChatGPT. From its earliest stages, ChatGPT has excelled in this area and
has become widely used. Wu et al. (2023) and Coyne & Sakaguchi (2023)
found that ChatGPT tends to make significantly more edits compared to
conventional AWCF systems like MS-NLP or Grammarly. As a result, the
corrected sentences rarely contain any remaining errors and are more fluent
and natural-sounding. However, the frequent edits sometimes lead to over-
corrections, with ChatGPT often making stylistic changes that are unnecessary for grammatically correct expressions. This can also alter the nuance of
the sentences, making it important for users to carefully review the output
text.

Figure 8.7 shows a comparison between the corrected sentences by Grammarly and those by ChatGPT. We used the free version of Grammarly and
OpenAI's GPT-3.5 Turbo model for ChatGPT.[7] The original input text is
as follows:

> I agree because is good idea the offer of anline or video and also thus can
> study from home. If students have work but they can study in home, it is
> not complicate , for students and thus they can improvement in the class.
> Today at day student can use tengnology (internet) and thus students can
> easy watch class.

In Figure 8.7, the substitutions and deletions were marked by striking
through the words, while the inserted words were highlighted in red. Chat-
GPT outputs were characterized by more edits, particularly word insertions,
as evidenced by the many words in red. Notably, the first sentence of Chat-
GPT output included four new words, 'option so that students,' which were
neither substitutions for any original words nor implied in the original text.

---

[7] We used a prompt including "Provide a grammatical correction for the input. Do not
change the meaning of the sentence or add any additional information. If the sentence
does not include any grammatical error, then return the sentence as it is." to generate
the grammatical error corrections from ChatGPT. We submitted the prompt and input
using the text completion API from `https://platform.openai.com/playground`.

Grammarly	ChatGPT
I agree because it is a good idea the offer ~~of~~ a online or video and also thus can study from home. If students have work but they can study ~~in~~ at home, it is not complicated, for students and thus they can improvement ~~in the~~ class. Today at day student can use ~~tengnology~~ technology (~~the~~ internet) and thus students can easy ~~ily~~ watch a class.	I agree because it is a good idea the ~~o~~ offer ~~of~~ an a online or video ~~and also thus~~ option so that students can study from home. If students have work but they can study ~~in~~ at home, it is not complicated, for students and thus they can improvement in the class. Today, ~~at day~~ students can use ~~tengnology~~ technology (internet) and thus students can easy ~~ily~~ watch class.

Fig. 8.7 Comparison between Grammarly and ChatGPT grammatical error corrections

The insertion of these words, along with additional edits, made the sentences much more natural compared to the output from Grammarly. However, they also made the output look meaningfully different from the original input. On the contrary, the output of Grammarly was more faithful to the original text, but the meaning of most sentences was still obscure.

Providing functions far beyond basic grammar corrections has become a popular trend for writing assistance systems; they now excel at making sentences more natural and adapting tone to fit any context. These advanced functions allow writers to rewrite entire paragraphs in various styles. Figure 8.8 shows an example of changing tones using ChatGPT. The input text is identical with Figure 8.7. The left column is a rewritten version in a 'persuasive' tone, while the right column is a rewritten version in an 'formal' tone.

Both texts were substantially different from the original sentences. The word overlap with the original sentences was under 50% for both texts, which was substantially lower than the 89% for grammatically corrected sentences by Grammarly and 85% for those corrected by ChatGPT. Although the general contents of these new texts remained consistent with the original texts, the characteristics and quality of the sentences were totally different.

These 'pain-free' rewriting functions make the boundaries between AWCF systems and other generative language models unclear. Recently, Grammarly also integrated LLMs into its system, and many new functions such as 'Sentence Rewriter' and 'Paragraph Rewriter' were added. These changes led some universities in the U.S. to place Grammarly under general AI bans. One such university was the University of North Georgia. The policy change created confusion among professors and students, as seen in Marley Stevens's case (Tang, 2024). She was accused of 'unintentional cheating' by her professor after revising her criminal justice essay using Grammarly. Her situation was even more perplexing because her university initially recommended

Persuasive	Formal
I ~~agree~~concur, as ~~because is good idea the~~ offering of a~~online~~ or video-based learning ~~and also~~ thus can ~~study from home.~~ is indeed a brilliant idea. ~~If~~This allows students ~~have work but they can~~ to study ~~in~~from the comfort of their homes~~,~~. Even if students are working, studying from home simplifies the process, ~~it is not complicate~~, ~~for students and thus they can~~ enabling them to ~~improvement in the~~ class. In ~~T~~today's world, ~~at day~~students can ~~use~~ leverage ~~tengnology~~ technology, particularly the (internet), ~~and thus students can~~ to easy~~ily watch~~ access their classes.	I ~~agree~~concur with the proposition ~~because~~as it is indeed a ~~good~~beneficial idea ~~the~~to offer ~~of~~ a~~online~~ or video-based learning, ~~and also~~ thus ~~can~~enabling students to study from home. If students ~~have work~~are employed, ~~but they can~~ study~~ing in~~ from home, ~~it is not~~ will not pose any complicate~~ions~~, and will facilitate their academic ~~for students and thus they can~~ improvement ~~in the class~~. ~~Today at day~~ In this moder era, student can ~~use~~utilize ~~tengnology~~ technology, specifically the ~~(internet) and thus students can~~ to easy~~ily~~ ~~watch~~access their classes.

Fig. 8.8 Examples of tone changes by ChatGPT

Grammarly on its website, then removed it, only to later reinstate it. This case highlights the challenge of establishing clear boundaries between plagiarism and AI-assisted work and the difficulty of developing a consistent guideline or policy. In summary, L2 writers need to be especially cautious when using these tools for academic purposes. Moreover, the changes in nuance or meaning that AI tools demonstrated in Figure 8.8 emphasize the importance of careful review. L2 writers must actively examine these revisions and decide whether to accept or reject them to ensure their intended meaning remains intact.

8.3.3.3 Use of MT for writing assistance

Unlike AWCF systems, which were generally accepted and even actively integrated into ESL/EFL writing courses, the role of MT in foreign language education has been widely debated. Many educators viewed MT as a threat to academic honesty and an obstacle to meaningful learning, as shown in the title of well-cited paper "Machine translation: friend or foe in the language classroom?" (Klekovkina & Denié-Higney, 2022) or "Friend or foe? Google Translate in language for academic purposes" (Groves & Mundt, 2015). The primary concern was academic dishonesty, with some educators viewing MT as a tool that could help students cheat. Another significant issue was the low accuracy of the translated sentences, which could expose students to large quantity of the faulty inputs. As a result of these and related concerns, several U.S. universities have banned the use of MT in graded written assignments within their language programs.

Regardless of their concerns, as the quality of free online MT services such as Google Translate™ has sharply improved, L2 writers' usage also has meaningfully increased. Based on the students' survey studies conducted between 2010 and 2020, MT usage has been consistently increased. Pena (2010)'s study, conducted in a secondary Spanish school in Australia, found around 33% students used MT. Later studies such as O'Neill (2019a) found that nearly nine out of ten students use MT even if its usage was prohibited within the institutions.

Clifford et al. (2013) conducted a survey study involving approximately 900 undergraduate students at one university in U.S. between 2011 and 2012. According to L2 writers' self-reported usage in the survey, the most frequent use of MT was for vocabulary (91%). Translating individual words was the most common usage, followed by translating short phrases. The survey also found that MT was used to write an entire sentence or multiple paragraphs, but the frequency of this usage was significantly lower. Some students also reported using MT to better understand instructions or to double-check their writing through back-translation. These reported usages were largely different from the usages educators were concerned about, and some of these usages might even facilitate language learning. The students' frequent and varied uses of MT, combined with low accuracy in detecting texts written with MT, raised questions about the prohibition of MT in the language programs. The policy may not be practical and might not necessarily benefit L2 learners.

Farzi (2016) and Chandra & Yuyun (2018) deepened our understanding of actual usage by recording computer screens during writing sessions and examining in-depth aspects such as frequencies and the goals of each use. They found that the actual usage was consistent with Clifford et al. (2013)'s survey: the main use of MT was as a replacement for a paper dictionary due to its convenience, such as efficiency in searching for words and ease of carrying. (Students already carry laptops, and with MT they do not need to carry an additional item, such as a paper dictionary.) However, as MT accuracy improves, its use for translating longer texts may increase, given that the primary reason students avoided translating longer texts was the low quality of MT for these units (Clifford et al., 2013). This raises further concerns about the potential misuse of MT in language education and underscores the importance of proper use by L2 learners.

Going beyond the analysis of students' usage, several studies have tried to identify better ways to use MT for L2 writers (Lee, 2020; Xu, 2022). Xu (2022) conducted a study where four Japanese learners at different proficiency levels used MT during the revision of the self-written essays. He introduced a multi-step revision process where L2 learners needed to write their first draft in L2 without MT and revise their draft using repetitive rounds of back-translation (from L2 to L1) and translation (from L1 to L2).

All students used these repetitive translation steps to verify whether their intentions were correctly expressed in L2, while the detailed patterns about how to integrate the output of MT into their own writing varied by proficiency levels. High-proficiency students mostly accepted word-level changes, while low and intermediate proficiency students accepted both word- and phrase-level changes. Notably, the writer who integrated the longest MT sentences into her work was the low-proficiency student. Unlike the other two learners, she made several sentence-level edits. Unfortunately, both the original translations and the modified sentences contained many errors, leaving the draft incorrect and unintelligible. This result was consistent with Kol et al. (2018)'s finding that low proficiency learners did not have an ability to identify and correct the errors in sentences generated by MT. As a result, the quality of the final draft for the low proficiency learner was not improved.

This result leads naturally to the next question, whether the output texts written with the assistance of MT are better or not. Farzi (2016), Kol et al. (2018) and O'Neill (2019b) found that L2 writers at the intermediate levels tended to write longer texts with a wider variety of grammatical expressions and vocabulary when they used MT. Similarly, Garcia & Pena (2011) and Xu (2022) found that beginners were able to produce longer texts with less effort, but the quality of the output was not improved. Despite these positive results in the output text, the impact on actual learning gains remains unclear. Only a few studies have investigated the relationship between MT and long-term learning gains, and they have reported negative results. For instance, O'Neill (2019b) analyzed essays written by 310 U.S. students enrolled in intermediate French and Spanish courses. The comparison between the groups with and without MT assistance showed that the former could write better essays; the holistic proficiency scores of the MT-assisted group were significantly higher than those of the group without MT. However, during the post-test, where all students wrote essays without MT, no meaningful differences were found. Similar results were observed in Fredholm (2019)'s experiment involving 31 Spanish learners. L2 writers achieved a higher degree of lexical diversity while using MT, but this effect vanished when they no longer used the tools. Based on these results, O'Neill (2019b) argued that the improvement observed with MT might come from familiarity with the tools rather than an enhancement of fundamental writing skills.

Still, it is possible for MT to support long-term learning gains for L2 learners. Valijärvi & Tarsoly (2019) showed that specially designed activities, such as correcting errors in sentences or texts translated by MT, can help draw learners' attention to language structure and boost their metalinguistic awareness. However, this approach requires careful preparation and specific guidance from teachers. For example, learning from translated sentences without knowing the correct forms can lead to confusion and even result in learning incorrect language.

Similar to AWCF systems, L2 writers may need to adopt dual approaches when incorporating MT in their writing process. For L2 writers who are writing for purposes other than learning, the focus should be on achieving high accuracy efficiently. Translating using MT followed by post-editing is a popular strategy due to its efficiency. While L2 writers use this strategy for their writing, they need to carefully review the MT output to ensure quality. For instance, understanding both common and L1-L2 pair-specific errors in MT is useful because it helps in identifying MT errors more effectively. The TextMaster, an online professional translation platform, provides a list of common MT error types and error examples. One of the examples they provided involved translation errors during the 2018 PyeongChang Olympics. The Norwegian team's chef received 15,000 eggs instead of the intended 1,500 due to a mistake that occurred while using Google Translate. Number translation errors are one of the well-known error types made by neural translation systems, and 1,500 eggs in the original text was translated as 15,000 eggs. Writers need to be aware of these issues and carefully review the translated outputs due to potentially significant impact (Dujardin, 2023).

On the other hand, as with other writing assistance tools, it is important to include active learning steps, such as checking grammar rules and understanding vocabulary nuances with additional resources, when L2 writers use MT for learning purposes. Simply post-editing MT output does not enhance language skills, as noted by O'Neill (2019a) and Fredholm (2019).

8.4 Summary

Despite significant improvements in NLP and AI, which have led to an improvement in accuracy and addition of various new features in writing assistance tools, there are still many issues with these systems. For researchers and developers who create these systems, it is important to recognize that achieving true learning gains and efficient writing may require different types of support, and systems should be designed to accommodate both goals.

Meanwhile, both L1 and L2 writers need to be actively engaged and capable of mitigating problems to use these tools effectively and appropriately. Recent transformative advancement of these tools has raised new concerns about academic honesty and writers' learning gains. However, detailed studies about how writers use these tools have demonstrated that there are many proper modes of usage that can actually expedite learning. Therefore, blindly prohibiting use of these tools for writers may be neither effective nor practical. Teachers and educational policy makers need to be actively involved and provide proper guidance and support for students in how to use these tools.

Chapter 9
Summary

9.1 Prelude: Putting pen to paper

The first inventors of writing were not particularly concerned with simplicity.

In Mesopotamia, the accounting system from which writing evolved was originally designed for another purpose—not to record language but to record commodities. As writing evolved, it became more and more sophisticated. First, in order to become writing in the strict sense at all, scribes had to discover the principle that one could write words not only with reference to what they meant, but also for how they sounded. Then, after this discovery, as more and more ways of representing sounds were added to the system, it also became possible to represent more and more fine-grained linguistic information. The latest Sumerian texts represented the language far more faithfully than the earliest texts. But as with any organic evolutionary development, material accreted but was less frequently lost, so that the mature system was considerably more complicated than the system it started from. The same was true for example in the case of Chinese, so that the script used in the Han dynasty was more complex than that of the original Shang Oracle Bone inscriptions—and the modern system more complex still.

Simplification in scripts was invariably a byproduct of borrowing across cultures, or of disruptive change within a culture. The borrowing of the idea of writing in the ancient Sinai where Egyptian writing inspired Semitic speakers to develop their own, much simpler, segmental system is an instance of the first. King Sejong's idea to make it easier for the common people to read and write, leading to the invention of Hangul, was an instance of the second.

But disruptive change does not always engender this result, and sometimes exactly the opposite could happen. When the Chinese system was adopted into Japanese, there was in principle an opportunity for simplification. The Japanese started early using Chinese characters for their sounds to represent native words, and in that usage Chinese characters evolved into the phonetic Kana scripts. Here there was an opportunity to abandon the original logographic aspects of Chinese writing entirely. Yet this path was not taken, and the result is the Modern Japanese writing system, which in many ways rivals in complexity the original cuneiform-based writing systems of Ancient Mesopotamia.

And even when a simplification is introduced, it is invariably guided by considerations of the here and now. Sejong was concerned with the number of symbols that the writer had to learn, and so settled on a compact phonetic script. The arrangement of symbols in that script as they were to appear on the page, however did not figure in the simplification process. Sejong could hardly have anticipated the difficulties for typing and printing technology that the arrangement of the *jamo* segments into syllables would cause.[1] Indeed, there was probably a strong motivation *not* to arrange the symbols linearly: Chinese characters, phonologically, represented syllables, so arranging *jamo* into syllable blocks would have felt more culturally familiar.

The Sinaitic scripts and many of their descendants, including the modern Latin alphabet, were technologically luckier. Egyptian writing, while not totally linear, was often quite close to linear, and in Sinaitic writing, the letters were arranged completely linearly. This carried over into later scripts—Phoenician, then Greek, then Latin, Cyrillic and so forth. When printing, and later typing came about, adapting the new technologies to these scripts was a piece of cake.

But not all descendants of the Sinaitic scripts were so lucky. The most plausible theory for the origins of Brahmic scripts is that they were inspired by, if not directly borrowed from, the Aramaic scripts, themselves descended from the Sinaitic scripts (Salomon, 1996). Yet the decision to represent vowels as diacritics rather than as inline letters—a decision that was neutral as far as writing with a pen or inscribing on stone were concerned—would prove problematic when printing and subsequently typing were introduced into India. Further complications arose as the daughter scripts were adapted to writing Sanskrit, which in general had more complex consonant clusters than its descendant, Pali, for which Brahmi had originally been developed. The resulting consonant ligatures, which in some of the descendant scripts became quite baroque in their complexity, further added to the difficulty of adapting printing and typing technology. Of course they also made the script more difficult to master for the scribe.

In English the act of writing is often referred to as the act of "putting pen to paper". This expression makes reference to three of the components that are involved in transferring one's thoughts, expressed in language, to a graphical form that can be read: The writing implement, in this case the pen; the writing surface, here the paper; and the "putter", i.e. the writer. The other two components are implicit: the script used for the language, and how that script represents the language.

---

[1] As we saw in Chapter 3, movable type was known in Korea from well before Sejong's time, but it was not widely used, the standard printing technology being woodblock printing, which is largely agnostic to how the symbols in a script are arranged.

We have discussed each of these components in some depth in this book, from the point of view of the technological complexities and user abilities that come into play when "putting pen to paper".

The script itself can present difficulties. While pens, styluses, brushes and such traditional writing implements are to a large extent agnostic about the complexity of the script, the script itself may nonetheless be complex to learn. But even if the script is easy, properties of how the elements of the script are laid out on the page can make matters difficult for technologies that came after pens, styluses and brushes. And while modern computing technology has to a large extent solved these sorts of problems, keyboards are still limiting factors when it comes to designing input systems for many of the world's scripts.

The writing system—how the language is represented in the script—can also be problematic, largely at the level of cognitive processing for writers (and readers). Traditional writing technology could do little to help with this. A pen could not help bad spellers improve their spelling, nor could a brush help a poor scholar in premodern China who could not remember how to write a given character. In such cases one would have to resort to auxiliary data in the form of dictionaries, which was effortful and time consuming. Once again, computer technology has narrowed the gap in that the writing implement, in becoming "intelligent"—or at least embodying a simulation of intelligence, can now present cogent hints to writers, correct their input, or even write for them.

The act of "putting pen to paper" necessarily involves gestures, whether those gestures involve manipulating a brush or pen, typing on a keyboard, or swiping across the surface of a smart screen. Such gestures vary greatly in the motor control involved. For some populations, in particular people with various sorts of motor disabilities, certain actions may be impossible, necessitating alternative input technologies.

Even for people without such disabilities, there are circumstances where space and time are at a premium, and one must make use of abbreviatory devices in order to speed up, or compress the text one is writing. Such space and time pressure is by no means restricted to the modern era but goes back to the earliest writing systems. The Sumerian aphorism quoted on the dedication page of this book makes this point very well:

> A scribe whose hand can keep up with the mouth is indeed a scribe.

In fact the earliest texts from both Mesopotamia and Egypt were replete with various forms of abbreviations. In the modern context, the issue comes up in many guises. In the choice of script: many speakers of Indian languages that use Brahmic scripts find it less time consuming to type using the Latin alphabet, and have the input system convert to the native script, rather than type in the native script directly, though this is technically possible. And for users of Augmentative and Alternative Communication, including those with motor disabilities, there are powerful incentives to find ways to save time in inputing text. Co-construction is all about finding the most efficient ways to help users with disabilities express themselves.

Then there is the issue of language itself. As we have seen, it is not unusual for people to be literate, but not in their native language. And it is not unusual for there to be more than one way to write a language, the various ways being separated by sociological criteria. After the invention of Hangul in the Joseon dynasty, literacy did indeed spread among the population, but the result was, for many centuries, two separate communities of writers. The literati continued to use Chinese-based writing, whereas it was the common people who used Hangul. Still, the need for the literati to communicate in writing with people who did not have access to a classical education, ensured that the system could not be ignored even by the literati. Which in turn put the Hangul script in good stead when, much later and for nationalist reasons, the decision was made to abandon the older Chinese-based writing system.

Finally millions of people around the world need to learn to write in another language. For better or for worse, for most of the non-English-speaking world, the most common choice is English. This has spawned a whole industry of writing aids for non-native speakers of English, and here again the recent advent of Large Language Models has helped immensely with this task.

9.2 Postlude: The future of writing

In 2020, Elon Musk, promoting his Neuralink startup, predicted that "human language will be obsolete in as little as five years" though we could still "do it for sentimental reasons." (Embury-Dennis, 2020). Five years on, language seems to still be very much with us, and in the meantime Musk's 2020 bit of prophecy may seem like one of his less outlandish proclamations.

The bizarre ravings of lunatic "visionaries" aside, it seems that language is not going anywhere, but what about writing? Graphical systems that replace writing, but have at the same time the same communicative coverage as writing (which is to say, as natural language) have been the stuff of utopianists, and dystopianists alike. From John Wilkins' 1668 "philosophical language" to Charles Bliss's attempt to create a universal graphical system of communication (Bliss, 1965), many have been motivated to try to replace written language with a system that they feel is more "logical" or more directly communicates meaning. Eco (1995) discusses many such cases, and see also the final chapter of Sproat (2023). Science fiction writers also take up the theme. In the dystopian future of Neal Stephenson's *The Diamond Age*, much of the population communicates not in regular writing, but in *mediaglyphics*, animated symbols not unlike today's animated stickers. The psychohistorians in Isaac Asimov's *Second Foundation* communicate with each other purely with a combination of subtle gestures and mathematical symbology. Yet, even outside of science fiction there has been speculation about whether written language is due to be replaced by, say, texts based on emoji; one of the authors participated in an interview on WNYC radio about a decade ago on precisely this theme (WNYC, 2014).

Let us at this point voice our scepticism about such ideas: one can certainly communicate a lot of information without recourse to traditional writing, but what becomes difficult is trying to communicate the full range of what one can communicate in language using only symbols that make no reference at all to any specific language. See again, Sproat (2023) for an extensive discussion of this issue. Even if one could replace writing with an emojified "written language", this would in any case not really get around most of the difficulties of "putting pen to paper". Certainly one might avoid some language-particular issues: a text written in emoji does not have to deal with the intricacies of English spelling, or the difficulties with deciding how to read a given Kanji in a Japanese text, issues that we have discussed at length. And inasmuch as emoji texts happen to be linearly arranged, there is no need for a machine, mechanical or computational, to deal with anything akin to the intricacies of how individual letters and diacritics are arranged into *akshara* in Brahmic scripts.

Still, such symbol systems typically comprise a relatively large set of symbols, which one must learn. Emoji number over 3,500 according to emojipedia.org, which is a sizable number, larger than the roughly 2,100 Kanji considered to be "common use" in Japanese. One can categorize them (smileys, animals, food, ...) but still if one wants to type an emoji text, one has to find the relevant symbols. Workaround input techniques such as typing :dog: for 🐕 rather beg the question of whether one can escape written language at all. The same issue arises for, say, Blissymbols (Bliss, 1965). While this system has been adapted for use by some communities of people with disabilities as a device for AAC, and while there have been other such ostensibly nonlinguistic graphical systems that have been specifically designed for that purpose—e.g., Minspeak®—input still requires the design of keyboards that allow the user to easily navigate and find the desired symbol. A user who, due to motor disabilities, is limited to pressing a single button, is not necessarily going to find it easier to type in Blissymbols than in English.

But in any case, in many ways the future of traditional writing has never looked brighter. A lot of the difficulties with adapting various scripts to technology that we discussed in Chapter 3 have been largely solved by modern 'smart' computer-based systems. So can it be said that many of the problems with writing in complex writing systems (Chapter 4) have, if not solved, at least been greatly helped by recent advantages in Artificial Intelligence. And the same goes for many of the other technical problems faced by those who would put "pen to paper" that we discussed in subsequent chapters. Even if one is not a native speaker of Japanese, one can put together a plausible Japanese document with the aid of automated translation systems or Large Language Models. Indeed, while the practice is more honored in the breach than in the observance, you can even cheat and let the LLM write the text for you given a simple prompt in your own language.

Charles Bliss had a strong humanitarian motivation for creating his symbol system. A refugee from Nazi Germany who escaped first to Shanghai, then ultimately to Sydney, Australia, he felt that the rise of Fascism, and other evils of the first half of the 20th century were due in large part to a failure of people across cultures and languages to communicate. Hence the need for a universal communication system, one that would be easy to learn, and whose symbols would wear their interpretation on their sleeves, without any reference to any particular language. Bliss was hardly the first to voice this sort of utopian ideal: L. L. Zamenhof had similar ideas when he created Esperanto a half century earlier. One can, in any case, hardly fault Bliss's motivations, even if his goals of providing a universal communication system were not achieved.

Today, as the exact same forces as plagued the world in Bliss's day are rearing their ugly heads across the globe, the needs for universal cross-cultural communication could hardly be more pressing. Fortunately computational tools exist now that greatly facilitate this. Today it is almost possible for nearly everyone to become totally monolingual, and rely on AI to help with communication. While we most definitely do *not* advocate taking such an approach—learning foreign languages and cultures is as important as it always has been—still, it is good to know that the tools are there if you need them. But this suggests both that human language (despite Musk's prediction) is still needed, and that written language is likely to be a part of the communication process. If nothing else, the great advances in AI over the past few years have been enabled in large measure by massive amounts of *written* text used in the systems' training. Rather than seeking to replace written language with what must inevitably be poor substitutes, we should be developing ways to harness AI as a force for good in helping people use one of civilization's earliest technological inventions.

References

ADLER, MICHAEL H. 1973. The Writing Machine: a History of the Typewriter. London: George Allen & Unwin.

ADLER, SIMON. August 14 2020. The Wubi effect: Radiolab. WNYC Studios. URL `https://radiolab.org/podcast/wubi-effect`.

ALGHAMDI, HAMDAH ABDULLAH. 2018. Arabizi: An exploration of the use of the contemporary youth netspeak on Social Networking Sites in Saudi Arabia. PhD thesis, University of Canberra.

ALHARBI, WAEL. 2023. AI in the foreign language classroom: A pedagogical overview of automated writing assistance tools. Education Research International, 2023.1–15.

ALSULAMI, ASHWAQ. 2019. A sociolinguistic analysis of the use of arabizi in social media among saudi arabians. International Journal of English Linguistics, 9(6).257–270.

ANSALDO, ANA INÉS; KARINE MARCOTTE; LILIAN SCHERER and GAELLE RABOYEAU. 2008. Language therapy and bilingual aphasia: Clinical implications of psycholinguistic and neuroimaging research. Journal of Neurolinguistics, 21(6).539–557.

ATANASIU, VLAD. 2025. Ugraphia: The Pursuit of Perfect Legibility. Number 6 in Grapholinguistics and its Applications, Nantes: Fluxus Editions.

ATLAS, STEPHEN. January 2023. ChatGPT for higher education and professional development: A guide to conversational AI. URL `https://digitalcommons.uri.edu/cba_facpubs/548`. Accessed: 2024-10-20.

BAGNALL, ROGER S and RAFFAELLA CRIBIORE. 2006. Women's Letters from Ancient Egypt, 300 BC-AD 800. Ann Arbor: University of Michigan Press.

BARASA, SANDRA. 2016. Spoken code-switching in written form? Manifestation of code-switching in computer mediated communication. Journal of Language Contact, 9(1).49–70.

BARROT, JESSIE S. 2023. Using automated written corrective feedback in the writing classrooms: Effects on L2 writing accuracy. Computer Assisted Language Learning, 36(4).584–607.

BAUBY, JEAN-DOMINIQUE. 1997. The Diving Bell and the Butterfly. Notting Hill: Fourth Estate. Translated by J. Leggatt.

BAUTISTA, MARIA LOURDES S. 2004. Tagalog-English code switching as a mode of discourse. Asia Pacific Education Review, 5.226–233.

BAXTER, WILLIAM. 1992. A Handbook of Old Chinese Phonology. Number 64 in Trends in Linguistics, Studies and Monographs, Berlin: Mouton de Gruyter.

BEECHING, WILFRED. 1990. The Century of the Typewriter. Leicester: British Typewriter Museum Publishing.

BELL, AARON. 2011. Mechanisation dead ends: Forgotten Hangeul of the Intertype Corporation, Mergenthaler Linotype and the Monotype Corporation Ltd. Master's thesis, University of Reading, Reading.

BERG, KRISTIAN and MARK ARONOFF. 2017. Self-organization in the spelling of English suffixes: the emergence of culture out of anarchy. Language, 93.37–64.

BESNER, DEREK and MARILYN CHAPNIK SMITH. 1992. Basic processes in reading: is the Orthographic Depth Hypothesis sinking? In FROST, RAM and LEONARD KATZ, editors, Orthography, Phonology, Morphology and Meaning, volume 94 of Advances in Psychology, chapter 3, pages 45–66. Amsterdam: North-Holland.

BEUKELMAN, DAVID R and PAT MIRENDA. 1998. Augmentative and alternative communication. Baltimore: Paul H. Brookes.

BI, YANCHAO; ZAIZHU HAN and YUMEI ZHANG. 2009. Reading does not depend on writing, even in Chinese. Neuropsychologia, 47(1193-1199).

BLACK, ROLF; JOSEPH REDDINGTON; EHUD REITER; NAVA TINTAREV and ANNALU WALLER. 2010. Using NLG and sensors to support personal narrative for children with complex communication needs. In Proceedings of the NAACL HLT 2010 Workshop on Speech and Language Processing for Assistive Technologies, pages 1–9. URL https://aclanthology.org/W10-1301.

BLICKENSDERFER MANUFACTURING COMPANY. 1908. Instructions for using the Blickensderfer typewriters. URL https://archive.org/details/InstructionsForUsingTheBlickensderferTypewriters.

BLISS, CHARLES. 1965. Semantography. Sydney: Semantography (Blissymbolics) Publications.

BLOEMSAAT, J. G. 2006. Repetitive Stain Injury. A case of mind and matter. PhD thesis, Radboud Universiteit, Nijmegen, The Netherlands.

BLOOMFIELD, LEONARD. 1933. Language. Chicago: University of Chicago.

BNC CONSORTIUM. 2007. British National Corpus, XML edition. URL http://hdl.handle.net/20.500.14106/2554. Literary and Linguistic Data Service.

BONAKER, NICHOLAS; EMLI-MARI NEL; KEITH VERTANEN and TAMARA BRODERICK. 2023. A usability study of Nomon: A flexible interface for single-switch users. In Proceedings of the 25th International ACM SIGACCESS Conference on Computers and Accessibility, pages 1–17.

BÖRNER, WOLFGANG. 1989. Planen und Problemlösen im fremdsprachlichen Schreibprozeß: Einige empirische Befunde. Variatio linguarum: Beiträge zu Sprachvergleich und Sprachentwicklung, pages 43–62.

BOSCH, ANTAL VAN DEN; ALAIN CONTENT; WALTER DAELEMANS and BEATRICE DE GELDER. September 1994. Analysing orthographic depth of different languages using data-oriented algorithms. In Proceedings of the 2nd International Conference on Quantitative Linguistics, Moscow.

BRAILLE, LOUIS. 1829. Procédé pour écrire les Paroles, la Musique et le Plain-chant au Moyen de Points à l'Usage des Aveugles et Disposé pour eux. Paris: Institution Royale des Jeunes Aveugles.

BRAILLE, LOUIS. 1839. Nouveau procédé pour représenter par des Points la Forme même des Lettres, les Cartes de Géographie, les Figures de

Géométrie, les Caractères de Musique, etc. à l'Usage des Aveugles. Paris: Institution Royale des Jeunes Aveugles.

BRANDT, CARMEN. 2020. From a symbol of colonial conquest to the scripta franca: The Roman script for South Asian languages. In Wege durchs Labyrinth: Festschrift zu Ehren von Rahul Peter Das, pages 1–36: CrossAsia-eBooks Heidelberg.

BRODERICK, TAMARA and DAVID JC MACKAY. 2009. Fast and flexible selection with a single switch. PloS one, 4(10).e7481.

BURSTEIN, JILL; NORBERT ELLIOT; BEATA BEIGMAN KLEBANOV; NITIN MADNANI; DIANE NAPOLITANO; MAXWELL SCHWARTZ; PATRICK HOUGHTON and HILLARY MOLLOY. 2018. Writing mentor: Writing progress using self-regulated writing support. Journal of Writing Analytics, 2.285–313.

CAI, SHANQING; SUBHASHINI VENUGOPALAN; KATIE SEAVER; XIANG XIAO; KATRIN TOMANEK; SRI JALASUTRAM; MEREDITH RINGEL MORRIS; SHAUN KANE; AJIT NARAYANAN; ROBERT L MACDONALD and OTHERS. 2024. Using large language models to accelerate communication for eye gaze typing users with ALS. Nature Communications, 15.9449.

CARRIER, L MARK and SANDRA Y BENITEZ. 2010. The effect of bilingualism on communication efficiency in text messages (SMS). Multilingua.

CAVE, RICHARD. 2024. How people living with amyotrophic lateral sclerosis use personalized automatic speech recognition technology to support communication. Journal of Speech, Language, and Hearing Research, pages 1–17.

CHANDRA, SYLVI OCTAVIANI and IGNASIA YUYUN. 2018. The use of Google Translate in EFL essay writing. LLT Journal: A Journal on Language and Language Teaching, 21(2).228–238.

CHARPIN, DOMINIQUE. 2023. Literacy in the old babylonian period. In PAULUS, SUSANNE, editor, Back to School in Babylonia, number 1 in ISAC Museum Publications, pages 59–65. Chicago: Institute for the Study of Ancient Cultures of the University of Chicago.

CHUNG, SOYEON. 2009. A comparative study of "Songs of Moon Illuminating on Thousand Rivers" and "Songs of Flying Dragons" in Korea – focus on diglossia. The Study of Korean Language and literature, 33.187–222.

CLEWS, JOHN. 1988. Language Automation Worldwide: The Development of Character Set Standards. British Library Research & Development Reports, Harrogate, UK: SESAME Computer Projects.

CLIFFORD, JOAN; LISA MERSCHEL and JOAN MUNNÉ. 2013. Surveying the landscape: What is the role of machine translation in language learning? @ tic. revista d'innovació educativa, (10).108–121.

CONPT. ND. 活字サイズの変遷と活版工程の機械化 (change in type sizes and the mechanization of the printing process). URL https://conpt.jp/kisekip.html?sn=8. 日本新聞製作技術懇話会 (Conference for Newspaper Production Technique–Japan).

COULMAS, FLORIAN. 1989. Writing Systems of the World. Oxford: Blackwell.

COULMAS, FLORIAN. 2003. Writing Systems: An Introduction to their Linguistic Analysis. Cambridge: Cambridge University Press.

COYNE, STEVEN and KEISUKE SAKAGUCHI. 2023. An analysis of GPT-3's performance in grammatical error correction. arXiv preprint arXiv:2303.14342.

CRYSTAL, DAVID. 2006. Language and the Internet. Cambridge: Cambridge University Press.

CUMMING, ALISTER. 1989. Writing expertise and second-language proficiency. Language learning, 39(1).81–135.

DANIELS, PETER. 1996. Analog and digitial writing. In DANIELS, PETER and WILLIAM BRIGHT, editors, The World's Writing Systems, pages 883–892. Oxford: Oxford University Press.

DANIELS, PETER. 2018. An Exploration of Writing. Sheffield: Equinox.

DANIELS, PETER and WILLIAM BRIGHT, editors. 1996. The World's Writing Systems. New York: Oxford.

DAVID, PAUL A. 1985. Clio and the economics of QWERTY. The American economic review, 75(2).332–337.

DEFRANCIS, JOHN. 1984. Digraphia. Word, 35(1).59–66.

DEFRANCIS, JOHN. 1989. Visible Speech: The Diverse Oneness of Writing Systems. Honolulu, HI: University of Hawaii Press.

DEMASCO, PATRICK; MATTHEW LILLARD and KATHLEEN MCCOY. 1989. Word compansion: allowing dynamic word abbreviations. In Proceedings of the 12th Annual RESNA Conference, pages 25–30.

DI DONATO, UMBERTO. 2019. Le Tastiere Magiche. Milano: Quattro.

DIKLI, SEMIRE. 2010. The nature of automated essay scoring feedback. Calico Journal, 28(1).99–134.

DIRINGER, DAVID. 1958. The Alphabet: A Key to the History of Mankind. New York: Hutchinson's Scientific and Technical Publications.

DIZON, GILBERT and JOHN MAURICE GAYED. 2021. Examining the impact of Grammarly on the quality of mobile L2 writing. JALT CALL Journal, 17(2).74–92.

DOS SANTOS, ALLESSANDRA ELISABETH; LARISA OLESOVA; CRISTIANE VICENTINI and LC DE OLIVEIRA. 2023. ChatGPT in ELT: Writing affordances and activities. TESOL Connections.

DRUCKER, JOHANNA. 1995. Alphabetic Labyrinth: The Letters in History and Imagination. London: Thames & Hudson.

DUJARDIN, AMANDINE. 2023. Machine translation: 10 examples of errors. URL https://www.textmaster.com/blog/example-machine-translation-errors.

DVORAK, AUGUST and WILLIAM L. DEALEY. May 1936. Typewriter keyboard. US Patent No. 2,040,248. URL https://patents.google.com/patent/US2040248A.

ECO, UMBERTO. 1995. The Search for the Perfect Language. Oxford: Blackwell. Translated by James Fentress.

ELLIS, NICK; MIWA NATSUME; KATRINA STAVROPOULOU; LORENC HOX-
    HALLARI; VICTOR VAN DAAL; NICOLETTA POLYZOE; MARIA-LOUISA
    TSIPA and MICHALIS PETALAS. 2004. The effects of orthographic depth
    on learning to read alphabetic, syllabic, and logographic scripts. Reading
    Research Quarterly, 39(4).438–468.
ELLIS, ROD. 2009. A typology of written corrective feedback types. ELT
    Journal, 63(2).97–107.
EMBURY-DENNIS, TOM.   May 9 2020.   Elon Musk predicts human
    language will be obsolete in as little as five years: 'We could
    still do it for sentimental reasons'.   The Independent.   URL
    https://www.independent.co.uk/tech/elon-musk-joe-rogan-
    podcast-language-neuralink-grimes-baby-a9506451.html.
EVANS, ARTHUR J. 1909. Scripta Minoa, the written documents of Minoan
    Crete, with special reference to the archives of Knossos. Oxford: Oxford:
    Clarendon Press.
FANG, TAO; SHU YANG; KAIXIN LAN; DEREK F WONG; JINPENG HU;
    LIDIA S CHAO and YUE ZHANG. 2023. Is ChatGPT a highly fluent
    grammatical error correction system? a comprehensive evaluation. arXiv
    preprint arXiv:2304.01746.
FARNELL, BRENDA. 1996. Movement notation systems. In DANIELS, PETER
    and WILLIAM BRIGHT, editors, The World's Writing Systems, pages 855–
    879. Oxford: Oxford University Press.
FARZI, REZA. 2016. Taming translation technology for L2 writing: Docu-
    menting the use of free online translation tools by ESL students in a writing
    course. PhD thesis, Université d'Ottawa/University of Ottawa.
FERGUSON, CHARLES. 1959. Diglossia. Word, pages 325–340. doi: https:
    //doi.org/10.1080/00437956.1959.11659702.
FERRANDO, IGNACIO. 2018. The adnominal linker -an in Andalusi Arabic,
    with special reference to the poetry of Ibn Quzmān (twelfth century). In
    HOLES, CLIVE, editor, Arabic Historical Dialectology: Linguistic and Soci-
    olinguistic Approaches, chapter 4, pages 96–111. Oxford: Oxford University
    Press.
FERRARA, SILVIA. 2022. The Greatest Invention: A History of the World in
    Nine Mysterious Scripts. New York: Macmillan.
FERRARA, SILVIA; LAURA TASSONI; BERND KROMER; LUKAS WACKER;
    MICHAEL FRIEDRICH; FRANCESCA TONINI; LORENZO LASTILLA;
    ROBERTA RAVANELLI and SAHRA TALAMO. 2024. The invention of
    writing on Rapa Nui (Easter Island). new radiocarbon dates on the Ron-
    gorongo script. Scientific Reports, 14(2794).1–9. https://doi.org/10.
    1038/s41598-024-53063-7.
FISCHER, STEPHEN ROGER. 1997. Glyphbreaker. New York: Springer.
FOWLE, ALI. 4 January 2017. The slow death of Myanmar's typewriter
    industry. https://www.bbc.com/news/av/world-asia-38452242.

FREDHOLM, KENT. 2019. Effects of Google Translate on lexical diversity: Vocabulary development among learners of Spanish as a foreign language. Revista Nebrija, 13(26).98–117.

GARCIA, IGNACIO and MARÍA ISABEL PENA. 2011. Machine translation-assisted language learning: writing for beginners. Computer Assisted Language Learning, 24(5).471–487.

GARRETT, KATHRYN L.; MARY BETH HAPP; JOHN M. COSTELLO and MELANIE FRIED-OKEN. 2007. AAC in the intensive care unit. In BEUKELMAN, D.R.; K.L. GARRETT and K.M. YORKSTON, editors, Augmentative Communication Strategies for Adults with Acute or Chronic Medical Conditions, pages 17–57. Baltimore, MD: Paul H. Brookes Publishing Co.

GEHANI, NARAIN. 1987. Document Formatting and Typesetting on the UNIX System. volume 1. Summit, NJ: Silicon Press.

GELB, IGNACE. 1952. A Study of Writing. Chicago: University of Chicago Press, 1st edition.

GELB, IGNACE. 1963. A Study of Writing: University of Chicago Press, 2nd edition.

GLASSNER, JEAN-JACQUES. 2000. Écrire à Sumer: L'invention du cunéiforme. Paris: Seuil.

GLASSNER, JEAN-JACQUES. 2003. The Invention of Cuneiform: Writing in Sumer. Baltimore: Johns Hopkins University Press. Translated by Zainab Bahrani and Marc van de Mieroop.

GNANADESIKAN, AMALIA. 2009. The Writing Revolution: Cuneiform to the Internet: Wiley.

GNANADESIKAN, AMALIA and RICHARD SPROAT. 2018. Writing systems. In Oxford Bibliographies: Oxford University Press. URL https://www.oxfordbibliographies.com/view/document/obo-9780199772810/obo-9780199772810-0221.xml.

GOBLE, CORBAN. August 1985. Mark Twain's nemesis: The Paige Compositor. In History Division at the Association for Education in Journalism and Mass Communication Annual Convention, pages 1–36, Memphis, TN.

GOH, EMA and HONGCHEN WU. 2024. Code-switching in computer-mediated communication by Gen Z Japanese Americans. Linguistics Vanguard, 10(1).615–628.

GOLDBERG, DAVID. January 1997. Unistrokes for computerized interpretation of handwriting. US Patent No. 5,596,656. URL https://patents.google.com/patent/US5596656.

GOLDING, ANDREW R. and DAN ROTH. 1998. A winnow-based approach to context-sensitive spelling correction. Machine Learning, 34.107–130. URL https://api.semanticscholar.org/CorpusID:12283016.

GONZALES, WILKINSON DANIEL WONG and YUEN MAN TSANG. 2023. The sociolinguistics of code-switching in Hong Kong's digital landscape: A mixed-methods exploration of Cantonese-English alternation patterns on WhatsApp. Journal of English and Applied Linguistics, 2(1).2.

GORMAN, KYLE and BRIAN ROARK. 2024. Abbreviation across the world's languages and scripts. In Proceedings of the Second Workshop on Computation and Written Language (CAWL)@ LREC-COLING 2024, pages 36–42.

GOVIL, HARI G. March 16 1937. Typographical font. US Patent 2,074,216.

GREEN, JORDAN R; ROBERT L MACDONALD; PAN-PAN JIANG; JULIE CATTIAU; RUS HEYWOOD; RICHARD CAVE; KATIE SEAVER; MARILYN A LADEWIG; JIMMY TOBIN; MICHAEL P BRENNER and OTHERS. 2021. Automatic speech recognition of disordered speech: Personalized models outperforming human listeners on short phrases. In Interspeech, volume 2021, pages 4778–4782.

GREGG, JOHN ROBERT. 1916. Gregg Shorthand: A Light-line Phonography for the Million. New York: Gregg Publishing Company, new and revised edition.

GROVES, MICHAEL and KLAUS MUNDT. 2015. Friend or foe? Google Translate in language for academic purposes. English for Specific Purposes, 37. 112–121.

HALFF, RAPHAEL. 2023. Adventures of a bad researcher: The mystery of the last Yiddish linotype. https://www.yiddishbookcenter.org/language-literature-culture/vault/adventures-bad-researcher-mystery-last-yiddish-linotype.

HALPERN, JACK. 2006. The Kodansha Kanji Learner's Dictionary: Revised and Expanded. New York: Kodansha America.

HAN, YOUNG GYUN. 2015a. How to use manuscripts for the description of Korean language history proposals for a methodological reconsideration. The Society of Korean Historical Linguistics, 20.167–191. doi: http://dx.doi.org/10.14727/khl.2015.20.167.

HAN, YOUNG GYUN. 2015b. On setting-up 'establishing stage of modern Korean' and subdivision of the era. The Association of Korean Cultural Studies, 70.63–108.

HAN, YOUNG GYUN. 2017. On the changes in the recognitions of the unification of the written and spoken language and their applications – from the view point of establishment of modern Korean mix-script. Language Facts and Perspectives, 41.299–325.

HANDEL, ZEV. 2019. Sinography: The Borrowing and Adaptation of the Chinese Script. Number 1 in Language, Writing and Literary Culture in the Sinographic Cosmopolis, Leiden: Brill.

HANNAS, WILLIAM. 1997. Asia's Orthographic Dilemma. Honolulu: University of Hawaii Press.

HAYES, JOHN R and LINDA S FLOWER. 1980. The dynamics of composing: Making plans and juggling constraints. Cognitive processes in writing, pages 31–50.

HELLSTEN, LARS; BRIAN ROARK; PRASOON GOYAL; CYRIL ALLAUZEN; FRANÇOISE BEAUFAYS; TOM OUYANG; MICHAEL RILEY and DAVID RYBACH. 2017. Transliterated mobile keyboard input via weighted finite-state transducers. In Proceedings of the 13th International Conference on Finite State Methods and Natural Language Processing (FSMNLP 2017), pages 10–19. URL https://aclanthology.org/W17-4002/.

HEO, CHUL. 2010. Examination how many using compound of Chinese character words and investigate the frequency of use by using analysis of Modern Korean words 1, 2. Hanmun Gyoyuk Yeongu, 34.221–244. Title in English: "Investigation of the Proportion of Sino-Korean Words and the Utilization of Chinese Characters through [Modern Korean Usage Frequency Survey 1 2]".

HILL, PATI. 1957. Truman Capote: The art of fiction no. 17. The Paris Review, 16.35–51.

HLÁDEK, DANIEL; JÁN STAŠ and MATÚŠ PLEVA. 2020. Survey of automatic spelling correction. Electronics, 9(1670).1–29.

HOLES, CLIVE. 2018. Introduction. In HOLES, CLIVE, editor, Arabic Historical Dialectology: Linguistic and Sociolinguistic Approaches, chapter 1, pages 1–28. Oxford: Oxford University Press.

HOSOM, JOHN-PAUL; TOM JAKOBS; ALLEN BAKER and SUSAN FAGER. 2010. Automatic speech recognition for assistive writing in speech supplemented word prediction. In Interspeech, volume 2010, pages 2674–2677.

HOWE, ELLIC. 1947. The London Compositor: Documents relating to Wages, Working Conditions and Customs of the London Printing Trade 1785–1900. London and Beccles: William Clowes and Sons, Ltd.

HUANG, SHUTING; YACONG ZHOU; MENGLIN DU; RUIMING WANG and ZHENGUANG G. CAI. 2021. Character amnesia in Chinese handwriting: a mega-study analysis. Language Sciences, 85.101383. doi: https://doi.org/10.1016/j.langsci.2021.101383. URL https://www.sciencedirect.com/science/article/pii/S0388000121000309.

HUFFMAN, DAVID A. 1952. A method for the construction of minimum-redundancy codes. Proceedings of the IRE, 40(9).1098–1101.

HYLAND, SABINE. January 15 2021. Iconic signs in a non-iconic writing system: Khipus with potatoes, feathers, figurines and other objects. Presented at INSCRIBE Workshop on "Invention of Writing: Production of Images and Language Notation". URL https://www.inscribercproject.com/workshop/session4.php.

INSTITUUT VOOR NEDERLANDSE LEXICOLOGIE. 1995. Woordenlijst Nederlandse Taal. The Hague: Sdu Uitgevers.

INTELLIGENT. Feb 2024. 4 in 10 college students are using ChatGPT on assignments. URL https://www.intelligent.com/4-in-10-college-students-are-using-chatgpt-on-assignments/.

JAY, RICKY. 1986. Learned pigs & fireproof women. New York: Villard.

JEFFREY, ROBIN. March 15 1997. Oriya: "identifying ...with newspapers". Economic and Political Weekly, pages 511–514.

JEFFREY, ROBIN. 2000. India's Newspaper Revolution: Capitalism, Politics and the Indian-Language Press 1977–99. New York: St. Martin's Press.

JELINEK, FREDERICK. 1998. Statistical methods for speech recognition: MIT press.

JEONG, BYEONG-SEOL. 2008. Hangeul and publication as media in late Chosun period. The Chin-Tan Society, 106.145–164. ISSN 1013-4719.

JINCHO, NOBUYUKI; GARY FENG and REIKO MAZUKA. 2014. Development of text reading in Japanese: an eye movement study. Reading and Writing, 27.1437–1465.

JOHNY, CIBU; LAWRENCE WOLF-SONKIN; ALEXANDER GUTKIN and BRIAN ROARK. 2021. Finite-state script normalization and processing utilities: The Nisaba Brahmic library. In Proceedings of the 16th Conference of the European Chapter of the Association for Computational Linguistics: System Demonstrations, pages 14–23.

JURAFSKY, DANIEL and JAMES H. MARTIN. 2008. Speech and language processing.

KARATAY, YASIN and LEYLA KARATAY. 2024. Automated writing evaluation use in second language classrooms: A research synthesis. System, page 103332.

KATZ, LEONARD and RAM FROST. 1992. The reading process is different for different orthographies: The Orthographic Depth Hypothesis. In FROST, RAM and LEONARD KATZ, editors, Orthography, Phonology, Morphology and Meaning, volume 94 of Advances in Psychology, chapter 4, pages 67–84. Amsterdam: North-Holland.

KAYE, ALAN. 1976. Chadian and Sudanese Arabic in the Light of Comparative Arabic Dialectology. Number 236 in Janua Linguarum, Studia Memoriae Nicolai van Wijk Dedicata, Series Practica, The Hague: Mouton.

KELSO, DAVID and GREGG VANDERHEIDEN. 1982. Ten-branch abbreviation expansion for greater efficiency in augmentative communication systems. In Proc. of the 5th Annual Conf. on Rehabilitation Engineering, page 3.

KERNIGHAN, MARK D.; KENNETH W. CHURCH and WILLIAM A. GALE. 1990. A spelling correction program based on a noisy channel model. In COLING 1990 Volume 2: Papers presented to the 13th International Conference on Computational Linguistics. URL https://aclanthology.org/C90-2036/.

KHALIL, SAUSSAN. 2012. The evolution of the Arabic language through online writing: The explosion of 2011. In BRISMES 2012 Graduate Conference Papers: Change and Continuity in the Arab World.

KIM, DONG SEOK. 1978. Korean 4-set ball typewriter. URL https://doi.org/10.8080/2019780001033. Korean Patent.

KIM, TAE-HO. 2023. 한글과 타자기 (Hangul and the Typewriter). Gyeongi Goyang: Yeoksa Bipyeongsa.

KLEKOVKINA, VERA and LAURENCE DENIÉ-HIGNEY. 2022. Machine translation: friend or foe in the language classroom? L2 Journal: An electronic refereed journal for foreign and second language educators, 14(1).

KNIGHT, EDWARD HENRY. 1877. Knight's American mechanical dictionary. Cambridge, MA: Riverside Press.

KOCH FAGER, SUSAN; MELANIE FRIED-OKEN; TOM JAKOBS and DAVID R BEUKELMAN. 2019. New and emerging access technologies for adults with complex communication needs and severe motor impairments: State of the science. Augmentative and Alternative Communication, 35(1).13–25.

KOL, SARA; MIRIAM SCHCOLNIK and ELANA SPECTOR-COHEN. 2018. Google Translate in academic writing courses? The EuroCALL Review, 26(2).50–57.

KOLTOVSKAIA, SVETLANA. 2020. Student engagement with automated written corrective feedback (AWCF) provided by Grammarly: A multiple case study. Assessing Writing, 44.100450.

KOLTOVSKAIA, SVETLANA. 2023. Postsecondary L2 writing teachers' use and perceptions of grammarly as a complement to their feedback. ReCALL, 35 (3).290–304.

KONG, PYUNG WOO. January 1953. Korean typewriter. US Patent.

KRISTENSSON, PER-OLA and SHUMIN ZHAI. 2004. SHARK^2: a large vocabulary shorthand writing system for pen-based computers. In Proceedings of the 17th annual ACM symposium on User interface software and technology, pages 43–52.

KUKICH, KAREN. 1992. Techniques for automatically correcting words in text. ACM Computing Survey, 24.377–439.

KUSHLER, CLIFF. 1998. AAC using a reduced keyboard. In Proceedings of the California State University, Northridge Center on Disabilities Technology and Persons with Disability Conference.

LEE, HYUN-BOK. 1982. Language policy in communist countries: The case of north korea. Journal of Social Sciences and Policy Studies, 4(1).223–254.

LEE, JEONGBOK. 2014. History and implications of the linguistic and cultural movements in North Korea. Paedalmal, 55.159–191.

LEE, SANGMIN-MICHELLE. 2020. The impact of using machine translation on EFL students' writing. Computer assisted language learning, 33(3). 157–175.

LEVINE, SARAH; SARAH W BECK; CHRIS MAH; LENA PHALEN and JAYLEN PITTMAN. 2024. How do students use ChatGPT as a writing support? Journal of Adolescent & Adult Literacy.

LIN, ANGEL. 2005. Gendered, bilingual communication practices: mobile text-messaging among Hong Kong college students. Fibreculture Journal.

LINGUALISM. February 27 2020. Do Arabs write in their dialects of Arabic? URL https://lingualism.com/arabic/do-arabs-write-in-their-dialects-of-arabic/. Accessed: 2025-03-05.

LIU, MEILU; LAWRENCE JUN ZHANG and CHRISTINE BIEBRICHER. 2024. Investigating students' cognitive processes in generative AI-assisted digital multimodal composing and traditional writing. Computers & Education, 211.104977.

LIVERMORE, BENJAMIN. July 1863. Device for hand-printing. US Patent No. 39,296. URL https://patents.google.com/patent/US39296A.

LUCAS, CHRISTOPHER. 1979. The scribal tablet-house in ancient Mesopotamia. History of Education Quarterly, 19(3).305–332.

MacKENZIE, I SCOTT and R WILLIAM SOUKOREFF. 2002. Text entry for mobile computing: Models and methods, theory and practice. Human–Computer Interaction, 17(2-3).147–198.

MAIR, VICTOR. July 2010. Character amnesia. Language Log. https://languagelog.ldc.upenn.edu/nll/?p=2473.

MALLI, KARTHIK. 2022. Devanagari—the makings of a national character. https://www.typotheque.com/articles/devanagari-the-makings-of-a-national-character.

MANCHON, ROSA MARIA; JULIO ROCA DE LARIOS and LIZ MURPHY. 2000. An approximation to the study of backtracking in L2 writing. Learning and Instruction, 10(1).13–35.

MANOHAR, KAVYA and SANTHOSH THOTTINGAL. June 13–15 2018. Malayalam orthographic reforms. impact on language and popular culture. In HARALAMBOUS, YANNIS, editor, Graphemics in the 21st Century, volume 1, pages 329–351, Brest.

MARJOU, XAVIER. June 2021. OTEANN: Estimating the transparency of orthographies with an artificial neural network. In Proceedings of the Third Workshop on Computational Typology and Multilingual NLP.

McCALL, F.H. 1898. The Linotype Manual, Giving detailed instructions of the proper adjustment and care of the Linotype: Kansas City. URL https://archive.org/details/McCallLinotypeManual1898.

McILROY, M. DOUGLAS. 1982. Development of a spelling list. IEEE Transactions on Communications, 30.91–99.

MELETIS, DIMITRIS and CHRISTA DÜRSCHEID. 2022. Writing Systems and their Use: An Overview of Grapholinguistics. Berlin: De Gruyter.

MELVILLE, HERMAN. 1853. Bartleby, the scrivener: A story of Wall-Street. New York: G. P. Putnam's Sons.

MERGENTHALER LINOTYPE CORPORATION. 1933. Keyboard Operation Devanagari Linotype: For composing Sanskrit, Hindi, Marathi, Gujarati & other vernaculars of India. Brooklyn, NY: Mergenthaler Linotype Corporation.

MERGENTHALER LINOTYPE CORPORATION. 1940. Linotype Machine Principles. Brooklyn, NY: Mergenthaler Linotype Corporation.

MERGENTHALER LINOTYPE CORPORATION. ND. Linotype Faces. Brooklyn, NY: Mergenthaler Linotype Corporation.

MILLER, PAUL; TEVHIDE KARGIN and BIRKAN GULDENOGLU. 2014. Differences in the reading of shallow and deep orthography: developmental evidence from Hebrew and Turkish readers. Journal of Research in Reading, 37(4).409–432.

MOHANAN, K. P. 1996. Malayalam writing. In DANIELS, PETER and WILLIAM BRIGHT, editors, The World's Writing Systems, pages 420–425. Oxford: Oxford University Press.

MOHRI, MEHRYAR; FERNANDO PEREIRA and MICHAEL RILEY. 2002. Weighted finite-state transducers in speech recognition. Computer Speech & Language, 16(1).69–88.

MONTEE, BARBARA B; RAYMOND G MILTENBERGER and DAVID WITTROCK. 1995. An experimental analysis of facilitated communication. Journal of Applied Behavior Analysis, 28(2).189–200.

MONTES-ALCALÁ, CECILIA. 2024. Bilingual texting in the age of emoji: Spanish-English code-switching in SMS. languages 9: 144.

MOON, WILLIAM. 1873. Light for the blind. London: Longmans & Company.

MOOREHOUSE, ALFRED. 1953. The Triumph of the Alphabet. New York: Henry Schuman.

MULLANEY, THOMAS. 2012. The moveable typewriter: How Chinese typists developed predictive text during the height of maoism. Technology and Culture, 53(4).777–814. URL http://www.jstor.org/stable/41682742.

MULLANEY, THOMAS. 2017. The Chinese Typewriter: A History. Cambridge, MA: MIT Press.

MULLANEY, THOMAS. May 2021. Meet the mystery woman who mastered IBM's 5,400-character Chinese typewriter. Fast Company. https://www.fastcompany.com/90635203/ibm-chinese-typewriter-lois-lew.

MURRAY, DONALD M. 1972. Teach writing as a process not product. Report No Pub Date, 18.

NAG, SONALI and MARGARET SNOWLING. January 2011. Reading in an alphasyllabary: Implications for a language universal theory of learning to read. Article in Scientific Studies of Reading.

NAIK, B.S. 1971. Typography of Devanagari. Bombay: Directorate of Languages, Government of Maharashtra.

NAKAMURA, KEIKO. 2020. Written codeswitching in electronic discourse:texting in Japanese-English bilinguals. Educational Studies, 62. 129–136.

NATIONAL INSTITUTE OF KOREAN LANGUAGE. n.d. Standard korean language dictionary. URL https://stdict.korean.go.kr/main/main.do. Accessed: 2025-01-18.

NEIJT, ANNEKE and ANNEKE NUNN. 1997. The recent history of Dutch orthography: Problems solved and created. Leuvense Bijdragen, 86.1–26.

OKEN, BARRY; UMUT ORHAN; BRIAN ROARK; DENIZ ERDOGMUS; ANDREW FOWLER; AIMEE MOONEY; BETTS PETERS; MEGHAN MILLER and MELANIE FRIED-OKEN. 2014. Brain–computer interface with language model–electroencephalography fusion for locked-in syndrome. Neurorehabilitation and neural repair, 28(4).387–394.

ORHAN, UMUT; KENNETH E HILD; DENIZ ERDOGMUS; BRIAN ROARK; BARRY OKEN and MELANIE FRIED-OKEN. 2012. RSVP keyboard: An

EEG based typing interface. In 2012 IEEE International Conference on Acoustics, Speech and Signal Processing (ICASSP), pages 645–648.

OSAKA, NAOYUKI. 1989. Eye fixation and saccade during kana and kanji text reading: Comparison of English and Japanese text processing. Bulletin of the Psychonomic Society, 27(6).548–550.

OSWALD, ALISON. 2019. Devanagari script for the Mergenthaler Linotype. https://invention.si.edu/devanagari-script-mergenthaler-linotype.

OUYANG, TOM; DAVID RYBACH; FRANÇOISE BEAUFAYS and MICHAEL RILEY. 2017. Mobile keyboard input decoding with finite-state transducers. arXiv preprint arXiv:1704.03987. URL https://arxiv.org/abs/1704.03987.

OVERMANN, KARENLEIGH. 2016. Beyond writing: The development of literacy in the Ancient Near East. Cambridge Archaeological Journal, 26(2). 285–303.

OVERMANN, KARENLEIGH. 2022. Early writing: A cognitive archaeological perspective on literacy and numeracy. Visible Language, 56(1).9–44.

O'NEILL, ERROL M. 2019a. Online translator, dictionary, and search engine use among L2 students. CALL-EJ: Computer-Assisted Language Learning–Electronic Journal, 20(1).154–177.

O'NEILL, ERROL M. 2019b. Training students to use online translators and dictionaries: The impact on second language writing scores. International Journal of Research Studies in Language Learning, 8(2).47–65.

PALFREYMAN, DAVID and MUHAMED AL KHALIL. 2003. "a funky language for teenzz to use:" representing Gulf Arabic in instant messaging. Journal of computer-mediated communication, 9(1).JCMC917.

PARHAMI, BEHROOZ. 2019. Evolutionary changes in Persian and Arabic scripts to accommodate the printing press, typewriting, and computerized word processing. In Proceedings of the TeX Users Group Confeerence, pages 1–8, Palo Alto, CA. TeX Users Group.

PARK, JI-HOON. 2011. 새 활자 시대 초기의 한글 활자에 대한 연구 (research into Hangul type in the early modern era). Musashino Art University. http://koreantypography.org/wp-content/uploads/thesis/kst_j3_4.pdf.

PARKINSON, RICHARD. 1999. Cracking Codes: The Rosetta Stone and Decipherment. Berkeley, CA: University of California Press.

PARSONS, LORRAINE. 2018. The story of the Tibetan typewriter. Unitas Fratrum, 77.71–80.

PAULUS, SUSANNE, editor. 2023. Back to School in Babylonia. Number 1 in ISAC Museum Publications, Chicago: Institute for the Study of Ancient Cultures of the University of Chicago.

PELLEGRINO, FRANÇOIS; CHRISTOPHE COUPÉ and EGIDIO MARSICO. 2011. A cross-language perspective on speech information rate. Language, pages 539–558.

PERNIER, LUIGI. 1909. Il Disco di Phaestos con caratteri pittografici. Ausonia, 3.255–302.

PETERSON, JAMES. 1980. Computer programs for detecting and correcting spelling errors. Communications of the Association for Computing Machinery, 23(12).676–687.

PIRINEN, TOMMI and KRISTER LINDÉN. 2014. State-of-the-art in weighted finite-state spell-checking. In Computational Linguistics and Intelligent Text Processing: Proceedings of the CICLing 2014, volume 8404 of Lecture Notes in Computer Science, pages 519–532. Heidelberg: Springer.

POPE, MAURICE. 1975. The Story of Decipherment: From Egyptian Hieroglyphs to Linear B. Hong Kong: Thames and Hudson.

POPE, MAURICE. 1999. The Story of Decipherment: From Egyptian Hieroglyphs to Maya Script. Hong Kong: Thames and Hudson, revised edition.

POPLACK, SHANA. 1980. Sometimes I'll start a sentence in Spanish y termino en Español: toward a typology of code-switching. Linguistics, 18.

POWELL, BARRY. 2009. Writing: Theory and History of the Technology of Civilization. Chichester: Wiley-Blackwell.

PRATT, JOHN. December 1866. Type-writing machine. Patent No. 3163. URL https://www.gov.uk/government/publications/patent-for-a-type-writing-machine-from-1866.

PRICE, GLANVILLE. 1971. The French language: Present and Past. London: Edward Arnold.

QI, DONALD. 1998. An inquiry into language-switching in second language composing processes. Canadian Modern Language Review, 54(3).413–435.

QUDAH, MAHMOUD ALI and OTHERS. 2017. A sociolinguistic study: Diglossia in social media. In Conference Proceedings. Innovation in Language Learning 2017.

QWEN TEAM. September 2024a. Qwen2.5: A party of foundation models. URL https://qwenlm.github.io/blog/qwen2.5/.

QWEN TEAM. November 2024b. QwQ: Reflect deeply on the boundaries of the unknown. URL https://qwenlm.github.io/blog/qwq-32b-preview/.

RANALLI, JIM. 2021. L2 student engagement with automated feedback on writing: Potential for learning and issues of trust. Journal of Second Language Writing, 52.100816.

RANALLI, JIM. 2022. Automated written corrective feedback: Error-correction performance and timing of delivery. Language Learning and Technology, 26(1).1–25.

RAYNER, MANNY; BETH ANN HOCKEY; PIERRETTE BOUILLON and OTHERS. 2006. Putting Linguistics into Speech Recognition: The Regulus Grammar Compiler. Palo Alto, CA: CSLI Press.

RICHARDS, G. TILGHMAN. 1964. The history and development of Typewriters. London: Her Majesty's Stationary Office.

ROARK, BRIAN; RUSSELL BECKLEY; CHRIS GIBBONS and MELANIE FRIED-OKEN. 2013. Huffman scanning: using language models within fixed-grid keyboard emulation. Computer speech & language, 27(6).1212–1234.

ROARK, BRIAN; MELANIE FRIED-OKEN and CHRIS GIBBONS. 2015. Huffman and linear scanning methods with statistical language models. Augmentative and Alternative Communication, 31(1).37–50.

ROBINSON, ANDREW. 2007. The Story of Writing: Alphabets, Hieroglyphs & Pictographs. London: Thames & Hudson, 2nd edition.

ROBINSON, ANDREW. 2009. Lost Languages: The Enigma of the World's Undeciphered Scripts. London and New York: Thames & Hudson.

ROBSON, ELEANOR. 2001. The tablet house: a scribal school in Old Babylonian Nippur. Revue d'Assyriologie et d'Archéologie orientale, 93(1).39–66.

ROGERS, HENRY. 2005. Writing Systems: A Linguistic Approach. Malden, MA: Blackwell.

ROSCOE, ROD D; LAURA K ALLEN; JENNIFER L WESTON; SCOTT A CROSSLEY and DANIELLE S MCNAMARA. 2014. The Writing Pal intelligent tutoring system: Usability testing and development. Computers and Composition, 34.39–59.

ROSCOE, ROD D; JOSHUA WILSON; ADAM C JOHNSON and CHRISTOPHER R MAYRA. 2017. Presentation, expectations, and experience: Sources of student perceptions of automated writing evaluation. Computers in Human Behavior, 70.207–221.

ROSS, FIONA. 1988. The Evolution of the Printed Bengali Character. PhD thesis, School of Oriental and African Studies, University of London, London.

ROSS, FIONA. 2021. Invisible hands: tracing the origins and development of the Linotype Devanagari digital fonts. Journal of the Printing Historical Society, 3(2).111–153.

RUOSS, EMANUEL and JULIANE SCHRÖTER. 2020. Schweizerdeutsch: Sprach und Identität von 1800 bis heute. Basel: Schwabe Verlag.

RUSSELL, ROBERT C. April 1918. Specification of letters. US Patent.

RUSSELL, ROBERT C. November 1922. Index. US Patent.

SALOMON, RICHARD. 1996. Brahmi and Kharoshthi. In DANIELS, PETER and WILLIAM BRIGHT, editors, The World's Writing Systems, pages 373–383. Oxford: Oxford University Press.

SAMPSON, GEOFFREY. 1985. Writing Systems. Stanford, CA: Stanford University Press.

SAMPSON, GEOFFREY. 2012. Writing Systems. Stanford, CA: Stanford University Press, 2nd edition.

SCHERRER-SCHAUB, CRISTINA. 2016. Printing versus manuscript: History or rhetoric? a short note inspired by Pelliot DIC. In DIEMBERGER, HILDEGARD; FRANZ-KARL EHRHARD and PETER KORNICKI, editors, Tibetan Printing: Comparisons, Continuities and Change, volume 8 of Brill's Tibetan Studies Library, chapter 20, pages 153–170. Leiden, Boston: Brill.

SEYMOUR, PHILIP; MIKKO ARO and JANE ERSKINE. 2003. Foundation literacy acquisition in European orthographies. British Journal of Psychology, 94(2).143–174.

SHAW, GRAHAM W. 1977. The Cuttack Mission Press and early Oriya printing. British Library Journal, pages 29–43.

SHIEBER, STUART M and RANI NELKEN. 2007. Abbreviated text input using language modeling. Natural Language Engineering, 13(2).165–183.

SINGH, ARVIND KUMAR. December 1986. Impact of writing materials on the evolution of Brāhmi script. In Studies in Indian History and Culture: Anthology of Papers Presented at the 10th Annual Conference of Indian History and Culture Society, pages 133–140, Varanasi. Banaras Hindu University.

SINGH, VAIBHAV. 2018. The machine in the colony: technology, politics, and the typography of Devanagari in the early years of mechanization. Philological Encounters, 3(4).

SINGH, VAIBHAV. 2021. Minding the machine: Technological change, typographic resistance and print journalism in pre-independence India. Südasien-Chronik — South Asia Chronicle, 11.121–140.

SINHA, R. MAHESH K. January–March 2009. A journey from Indian scripts processing to Indian language processing. IEEE Annals of the History of Computing, pages 8–31.

SKJAERVO, P. OKTOR. 1996. Aramaic scripts for Iranian languages. In DANIELS, PETER and WILLIAM BRIGHT, editors, The World's Writing Systems, pages 515–535. Oxford: Oxford University Press.

SPROAT, RICHARD. September 1990. Method for conversion of phonetic Chinese to character Chinese. US Patent.

SPROAT, RICHARD. 1992. Morphology and Computation. Cambridge, MA: MIT Press.

SPROAT, RICHARD. 2000. A Computational Theory of Writing Systems. Cambridge: Cambridge University Press.

SPROAT, RICHARD. 2010. Language, Technology, and Society. Oxford: Oxford University Press.

SPROAT, RICHARD. 2023. Symbols: An Evolutionary History from the Stone Age to the Future. Cham, Switzerland: Springer Nature.

SPROAT, RICHARD and ALEXANDER GUTKIN. 2021. The taxonomy of writing systems: How to measure how logographic a system is. Computational Linguistics, 47(3).1–52.

SRIVASTAVA, ABHISHEK; KALIKA BALI and MONOJIT CHOUDHURY. 2020. Understanding script-mixing: A case study of hindi-english bilingual twitter users. In Proceedings of the 4th Workshop on Computational Approaches to Code Switching, pages 36–44.

STRONG, EARL POE. 1956. A Comparative Experiment in Simplified Keyboard Retraining and Standard Keyboard Supplementary Training. Washington, D.C.: General Services Administration. URL https://www.google.com/books/edition/A_Comparative_Experiment_in_Simplified_K/ZYXgAAAAMAAJ.

STUM, GREGG M and PATRICK DEMASCO. 1992. Flexible abbreviation expansion. In Proceedings of the RESNA International, volume 92, pages 371–373.

SU, YANFANG; YUN LIN and CHUN LAI. 2023. Collaborating with ChatGPT in argumentative writing classrooms. Assessing Writing, 57.100752.

SWATCH GROUP. 2012. Swatch Group Gschäftsbricht 2012, Eimaligi Dialäkt Usgaab!

TANG, WILLIAM. 2024. She used Grammarly to proofread her paper. Now she's accused of 'unintentionally cheating'. USAToday. URL https://www.usatoday.com/story/opinion/voices/2024/04/17/ai-students-cheating-plagiarism-grammarly/73223779007.

THI, NANG KHAM; MARIANNE NIKOLOV and KRISZTIÁN SIMON. 2023. Higher-proficiency students' engagement with and uptake of teacher and grammarly feedback in an EFL writing course. Innovation in Language Learning and Teaching, 17(3).690–705.

TIMMERS, SHANNON. 1987. The print medium in distance education. Technical report, Open Learning Institute/Knowledge Network, Vancouver, BC. URL https://dr.ntu.edu.sg/bitstream/10356/85946/1/AMIC_JUN8-11_1987_26.pdf.

TOMANEK, KATRIN; KATIE SEAVER; PAN-PAN JIANG; RICHARD CAVE; LAUREN HARRELL and JORDAN R GREEN. 2023. An analysis of degenerating speech due to progressive dysarthria on ASR performance. In IEEE International Conference on Acoustics, Speech and Signal Processing (ICASSP), pages 1–5. IEEE.

TOUTANOVA, KRISTINA and ROBERT C. MOORE. 2002. Pronunciation modeling for improved spelling correction. In Proceedings of the 40th Annual Meeting of the Association for Computational Linguistics, pages 144–151.

TSIEN, TSUEN-HSUIN. 1985. Paper and printing. In NEEDHAM, JOSEPH, editor, Science and Civilization in China, volume 5, pages 883–892. Cambridge: Cambridge University Press.

UNESCO. 2017. Reading the past, writing the future: Fifty years of promoting literacy. Technical report, United Nations Educational, Scientific and Cultural Organization, Paris.

URTON, GARY. 1998. From knots to narratives: Reconstructing the art of historical record keeping in the Andes from Spanish transcriptions of Inka Khipus. Ethnohistory, 45(3).409–438.

URTON, GARY. 2010. Numeral graphic pluralism in the colonial Andes. Ethnohistory, 57(1).135–164.

URTON, GARY. 2017. Inka History in Knots: Reading Khipus as Primary Sources. Austin, TX: University of Texas Press.

VALIJÄRVI, RIITTA-LIISA and ESZTER TARSOLY. 2019. 'language students as critical users of Google Translate': Pitfalls and possibilities. Practitioner research in higher education, 12(1).61–74.

VAN DER WEEL, F. .R. (RUUD) and AUDREY L.H. VAN DER MEER. 2024. Handwriting but not typewriting leads to widespread brain connectivity: a high-density EEG study with implications for the classroom. Frontiers in Psychology, 14.1–9. URL https://doi.org/10.3389/fpsyg.2023.1219945.

VANDERHEIDEN, GREGG and DAVID KELSO. 1987. Comparative analysis of fixed-vocabulary communication acceleration techniques. Augmentative and Alternative Communication, 3(4).196–206.

VOGT, HANS. 1954. Language contacts. Word, 10(2-3).365–374.

WARD, DAVID J; ALAN F BLACKWELL and DAVID JC MACKAY. 2000. Dasher—a data entry interface using continuous gestures and language models. In Proceedings of the 13th annual ACM symposium on User interface software and technology, pages 129–137.

WAŻNY, TOMASZ. 2016. Wooden book-covers, printing blocks, their identification and dating – how to read the wood. In DIEMBERGER, HILDEGARD; FRANZ-KARL EHRHARD and PETER KORNICKI, editors, Tibetan Printing: Comparisons, Continuities and Change, volume 39 of Brill's Tibetan Studies Library, chapter 20, pages 471–484. Leiden, Boston: Brill.

WILEY, TERRENCE G. 2017. Chinese "dialect" speakers as heritage language learners: A case study 1. In Heritage language education, pages 91–106: Routledge.

WILKINS, JOHN. 1668. An Essay Towards a Real Character, and a Philosophical Language. London: Royal Society.

WILSON, DOUG. 2012. Linotype: The film. URL https://linotypefilm.com/about.

WNYC. February 24 2014. The man who tried to eliminate all words, but never met a smartphone. Podcast. URL https://www.wnycstudios.org/podcasts/notetoself/articles/charles-bliss-emoji.

WOLF-SONKIN, LAWRENCE; VLAD SCHOGOL; BRIAN ROARK and MICHAEL RILEY. 2019. Latin script keyboards for South Asian languages with finite-state normalization. In Proceedings of the 14th International Conference on Finite-State Methods and Natural Language Processing, pages 108–117. URL https://aclanthology.org/W19-3114/.

WOODALL, BILLY R. 2002. Language-switching: Using the first language while writing in a second language. Journal of Second Language Writing, 11(1).7–28.

WOODS, CHRISTOPHER; EMILY TEETER and GEOFF EMBERLING, editors. 2010. Visible Language: Inventions of Writing in the Ancient Middle East and Beyond. Number 32 in Oriental Institute Museum Publications, Chicago: Oriental Institute.

WU, HAORAN; WENXUAN WANG; YUXUAN WAN; WENXIANG JIAO and MICHAEL LYU. 2023. ChatGPT or Grammarly? Evaluating ChatGPT on grammatical error correction benchmark. arXiv preprint arXiv:2303.13648.

XU, JUN. 2022. Proficiency and the use of machine translation: A case study of four japanese learners. L2 Journal, 14(1).

YAGHAN, MOHAMMAD ALI. 2008. "Arabizi": A contemporary style of Arabic slang. Design issues, 24(2).39–52.

YAN, DA. 2023. Impact of ChatGPT on learners in a L2 writing practicum: An exploratory investigation. Education and Information Technologies, 28 (11).13943–13967.

YANCEY, KEVIN P; GEOFFREY LAFLAIR; ANTHONY VERARDI and JILL BURSTEIN. 2023. Rating short L2 essays on the CEFR scale with GPT-4. In Proceedings of the 18th Workshop on Innovative Use of NLP for Building Educational Applications (BEA 2023), pages 576–584.

YOON, SU-YOUN; EVA MISZOGLAD and LISA R PIERCE. 2023. Evaluation of ChatGPT feedback on ELL writers' coherence and cohesion. arXiv preprint arXiv:2310.06505.

ZHANG, DE. 2004. Exploring the affordances technology provides for heritage learners of Chinese as a foreign language (cfl). In Society for Information Technology & Teacher Education International Conference, pages 4010–4017. Association for the Advancement of Computing in Education (AACE).

ZIMMERMANN, RÜDIGER. 2000. L2 writing: Subprocesses, a model of formulating and empirical findings. Learning and instruction, 10(1).73–99.

Index

abbreviation, 154–157, 164
  acronyms and initialisms, 156
  conventional vs. ad hoc, 155
  expansion, 166–167, 184
abjads, 25
abugida, 10, 25–26, 34, 45
  not to be confused with
    syllabaries, 26
Akhand system, 57–58
Akkadian, 43
akshara, 36, 60, 80
Alba, Thea, 152
alphabets, 25
alphasyllabary, 10, *see* abugida
ambiguous keyboards
  12-key layout, 139–141
  multitap, 140
  T9, 141–142
  two-key method, 141
amyotrophic lateral sclerosis
  (ALS), 23
Apollinaire, Guillaume, 43
Arabic, 25
  diglossia, 188
Arabic script
  for Turkish, 18
Arabizi, 191–192
Aramaic, 236
  script used for Middle Persian,
    90

arithmetic coding, 179–181
Armenian, 25
ASCII, 79, 135
Asimov, Isaac, 238
Augmentative and Alternative
    Communication, 7, 144–149,
    154, 237
  low-tech and high-tech,
    144–145
auto-correction, 164, 165, 173
automated writing corrective
    feedback system (AWCF),
    216, 217, 221–230
automated writing evaluation
    system (AWE), 216–217
automatic speech recognition,
    114, 116–117
  restricted vocabulary, 137–138

Bangla, 26
Banks, Iain M., 165
Basic Multilingual Plane, 80
Bauby, Jean-Dominique, 13–16,
    109, 112
Big5, 80
binary code, 16, 112, 175
binary switch, 109, 111
Blackletter, *see* Gothic type
Bliss, Charles, 238, 240
Blissymbols, 239

© The Editor(s) (if applicable) and The Author(s), under exclusive license
to Springer Nature Switzerland AG 2025
B. Roark et al., *Tools of the Scribe*,
https://doi.org/10.1007/978-3-032-00831-2

Author Index